GUIDE TO THE SELECTION OF ROTODYNAMIC PUMPS

Prepared by

the Engineering Equipment Users Association

GUIDE TO THE SELECTION OF ROTODYNAMIC PUMPS

E.E.U.A. Handbook No. 30: 1972

Published for

THE ENGINEERING EQUIPMENT USERS ASSOCIATION
20 Grosvenor Place, London, S.W.1

by

CONSTABLE & COMPANY LIMITED
LONDON, W.C.2

LONDON
Published in 1972 by
Constable & Co. Ltd.
10 Orange Street, W.C.2

ISBN 09 456820 0

Printed in Great Britain by
The Anchor Press Ltd., Tiptree, Essex

THE ENGINEERING EQUIPMENT USERS ASSOCIATION

The Engineering Equipment Users Association is an association of industrial users of engineering equipment, material and stores. It was founded in 1949 by a number of large companies who realised the important role that would be played in the future by the further standardisation and simplification of engineering products, and by the free exchange of technical information on engineering matters.

The principal objects of E.E.U.A. are to assist its members to promote a common policy for the standardisation of engineering materials, equipment and stores through the British Standards Institution or otherwise; to foster the preparation and adoption of national standards for such products and to collate and summarise information in order to give guidance on their nature and use. Membership is restricted to companies and other bodies who are predominantly users rather than manufacturers of engineering products.

The work of the Association is entirely complementary to that of the B.S.I. with whom close liaison and collaboration are always maintained; it does not duplicate the activity of that Institution in any way. Everything possible is done to enlist the early support of manufacturing interests in standardisation work which E.E.U.A. considers should be promoted in the national interest.

The Handbook was prepared by a panel drawn from member companies primarily to assist project engineers and designers in the selection, installation, and operation of rotodynamic pumps; bearing in mind the many factors which must be considered if the best equipment at the most economical cost, both capital and running is to be obtained.

The Council of E.E.U.A. decided that the work of this Panel should be made known outside the Association, and consequently the Handbook has been published for general sale.

ACKNOWLEDGEMENTS

The Association, as well as those who served on E.E.U.A. Panel M/39, wish to thank the following for information supplied during the preparation of the Handbook and for permission to make use of and reproduce certain data and illustrations.

Allen Gwynnes Pumps Limited.

American Institute of Chemical Engineers.

British Hydromechanics Research Association.

British Standards Institution.

Crane Limited.

Drysdale Pumps.

Hayward Tyler & Co Ltd.

Her Majesty's Stationery Office.

Institution of Water Engineers.

McGraw Hill Publishing Co. Ltd.

Pergamon Press Ltd.

Sigmund Pulsometer Pumps Limited.

Sulzer Brothers (London) Limited.

Trade and Technical Press Limited.

United Centrifugal Pumps (California).

Wydawictiura Naukowo—Techniczne.

Worthington Simpson Limited.

ENGINEERING EQUIPMENT USERS ASSOCIATION

GUIDE TO THE SELECTION OF ROTODYNAMIC PUMPS

E.E.U.A. Handbook No. 30 (1972)

This Handbook was prepared by E.E.U.A. Panel M/39 constituted from representatives of the following member organisations.

Central Electricity Generating Board

Courtaulds Engineering Limited

Shell International Petroleum Company Limited

The Distillers Company Limited

United Kingdom Atomic Energy Authority

GUIDE TO THE SELECTION OF ROTODYNAMIC PUMPS

(E.E.U.A. HANDBOOK No. 30: 1972)

CONTENTS

SYMBOLS USED

Symbol	Item	Units (length L; mass M; time T)
C	pipe friction constant	dimensionless
D	pipe internal diameter	L
d	impeller diameter	L
E	overall efficiency	%
f	friction factor	dimensionless
g	gravitational acceleration	L/T^2
H	total head available or required	L
h	static head	L
h_a	atmospheric pressure head $=p_a/w$	L
h_d	static head, delivery side	L
h_f	frictional head	L
h_i	impeller head loss	L
h_p	pressure head $=p/w$	L
h_r	residual head	L
h_s	static head, suction side	L
h_v	velocity head $=v^2/2g$	L
h_{vp}	head due to vapour pressure	L
μ	absolute viscosity	M/LT
N	pump speed	revolutions/T
$NPSH$	net positive suction head	L
n_s	specific speed	see Section Two **(b)**(ii)
γ	kinematic viscosity $=\mu/w$	L^2/T
P	power requirement q	ML/T
p	pressure	M/L^2
p_a	atmospheric pressure	M/L^2
p_T	pressure at syphon top	M/L^2
Q,q	flow rate	M/T or L^3/T
R_e	Reynolds Number $=\dfrac{vDw}{\mu}=\dfrac{vD}{\gamma}$	dimensionless
$S.G.$	specific gravity	M/L^3
v	velocity	L/T
w	specific weight of fluid	M/L^3
η	efficiency	%

THE SELECTION OF ROTODYNAMIC PUMPS

FOREWORD

This Handbook has been prepared to give some guidance to those responsible for the selection, installation and operation of rotodynamic pumps. It is primarily compiled for design and project engineers who may have a limited knowledge of pumping equipment, to assist them in selecting suitable equipment for specific applications.

At present, many types of pump, with widely differing characteristics, are used industrially, even for similar or identical applications: cost and performance thus vary correspondingly. Attention is drawn to the many factors that must be given consideration, if the best equipment at the most economical cost (both capital and running) is to be obtained.

Sections on basic theory to cover head and flow conditions, pump performance, speed, power, suction conditions, parallel and series operation, viscosity and surge effects are followed by sections on economic considerations, design and usage, installation, testing and commissioning, so that the Handbook is a general guide to the selection, installation and operation of pumping equipment.

Rotodynamic pumps, as envisaged in this book, are those pumps that impart pressure energy to the liquid being pumped, solely by the rotary action of the pump impeller. At any particular speed, substantial changes occur in the pressure or quantity parameters as one or the other is varied.

Other pumps, though rotary, such as the screw pump, gear pump, etc., of the positive displacement type (in which for any particular speed, the quantity pumped is largely independent of the pressure) are omitted, since the principle of operation is different.

The Handbook confines itself to the three main classifications of pump falling into a true *rotodynamic specification:* axial flow, centrifugal and mixed flow.

A consideration of the theoretical principles involved in the generation of pressure and discharge characteristics within the pump was deemed to be outside the scope, partly due to the desire to accentuate the 'selection' aspects, and partly because the subject has been widely covered elsewhere.

The design of the system in which the pump is to be used is considered in reasonable detail, however, for the logical approach to a suitable pump is to define clearly what it has to do, consider the circumstances in which it has to be done, and then apply the known characteristics of particular pump types and drivers to the requirements so defined. Sufficient information on the peripheral, vortex or regenerative pump is included to indicate its general classification and application.

The information given is based mainly on the experience of member organisations; it does not necessarily reflect the views, or embody the claims, of the suppliers and manufacturers of any equipment. Any reference to equipment which may be associated with particular manufacturers or suppliers does not imply that satisfactory equipment of a similar nature cannot be obtained from other sources or is not being developed elsewhere. Readers of the Handbook are in no way restricted from dealing with any particular manufacturer or supplier.

Wherever practical, SI units are used throughout, although certain tables, charts and nomograms contain alternative units, since basic data is still more readily available to many in the alternative units concerned. Duality of units in those circumstances was considered a reasonable compromise.

The Association wishes to thank all those who have supplied data or illustrations. The illustrations are intended to depict the basic features of design and do not purport to indicate details of design or of assembly.

A Bibliography and some useful conversion data appear at the end.

SECTION ONE

THE SYSTEM

(a) Head Conditions

Bernoulli's theorem, applied to an ideal fluid, expresses the hydraulic application of the law of conservation of energy. A liquid particle may have potential energy due to its position relative to a known datum, or pressure energy due to an applied pressure in its containment, also relative to the same datum. Or it may have kinetic energy at the same point. As motion occurs in the channel or pipeline containing the fluid, the energy at any two points is the same, allowing for losses due to friction in the pipe or channel.

This condition is represented diagramatically in Fig. 1.1 and, according to Bernoulli:

$$h_p + h_v + h_f + h = \text{constant}$$

Reference to Fig. 1.1 shows that the p/w factor determines the height

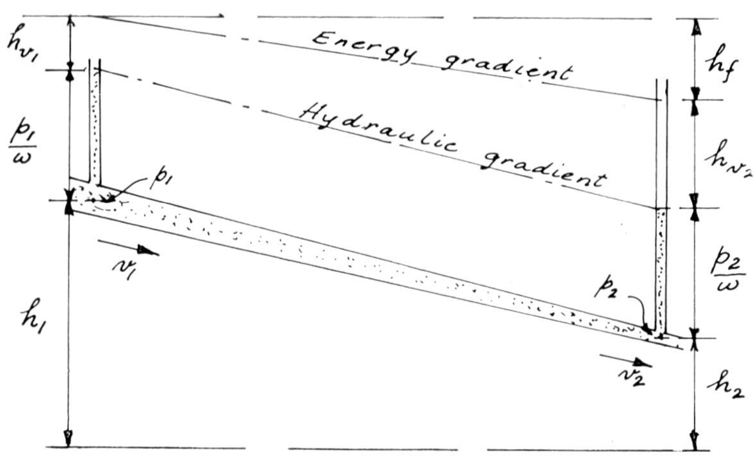

Fig. 1.1 Bernoulli's Theorem.

above the pipe to which the water would rise in an open-ended tube, that is, the atmospheric pressure point in each case. The line joining such points of atmospheric pressure is called the **hydraulic gradient**.

The total energy in the fluid at the same points relative to the pipe is, however, increased by the velocity head in each case, and by adding the $\frac{v^2}{2g}$ components to the pressure components, a further line is obtained depicting the **energy gradient**.

Finally, the difference between the energy gradient at one point and that at another is the measure of head loss (h_f) due to friction and other causes in the system.

It follows, then, that a liquid at high level uses its potential energy to propel itself under gravity along the connecting pipe. Conversely, where the liquid is to be raised to a higher level, an energy contribution is necessary from some external source, in order to overcome the frictional and other losses, and perform the necessary lifting action concerned.

Before any estimate can be made of the energy to be contributed to (or derived from) a system of fluid transfer, it is necessary to examine the basic types of system involved. Although complicated combinations are possible, and do occur, certain basic arrangements will serve to illustrate the essential design factors to be taken into consideration:

(i) *Gravitational systems*

In this instance (Fig. 1.2), water flows from an elevated point A to a lower point B. If free flowing, all the energy of the system is changed

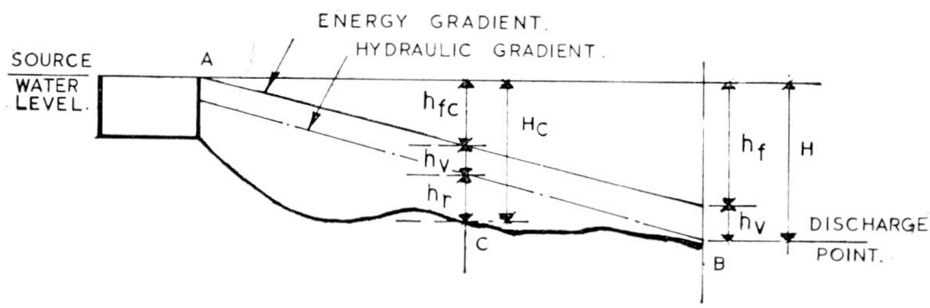

Fig. 1.2 Flow under gravity.

from potential energy at A to a terminal kinetic energy at B, after losses due to resistance in the system have been overcome, in which case we have:

$$H = h = h_f + h_v$$

At any intermediate point C, where the available head is H_c, not all of H_c will have been used up in overcoming friction and velocity generation. There remains a pressure head in the pipe at that point, therefore, referred to as the residual head h_r, which forms part of the head balance equations.

Thus, $$H_c = h_{fc} + h_v + h_r$$

is the general equation for balance of pressure conditions at any point of the pipe.

It should also be noted that the line drawn above the two residual components h_v and h_r represents the energy gradient of the system, whilst that drawn above h_r represents the hydraulic gradient.

(ii) *Assisted gravitational*

If, in the above example, the quantity of water finally arriving under gravity at B were insufficient for the purposes required, an additional energy contribution in the form of a pump would be required, say, at A. This pump would contribute to the system a head which would make up the deficit in h_v and the equation covering the Head Required (Fig. 1.3) is:

$$H = h_f + h_v - h$$

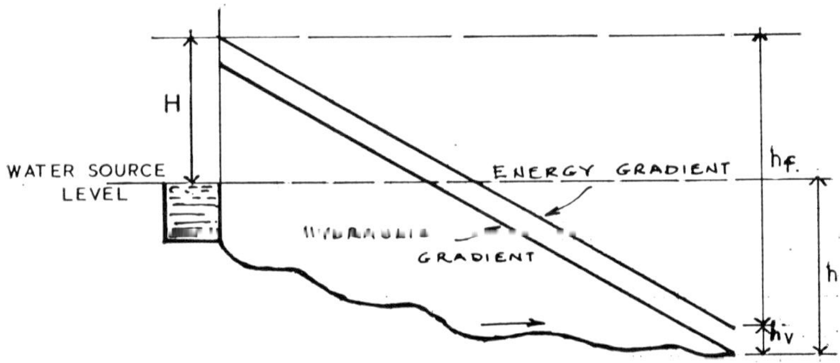

Fig. 1.3 Assisted gravitation.

(iii) *Elevated discharge systems*

Here gravitation opposes the direction of flow, and all the energy contribution to maintain flow has to be derived from some external source (Fig. 1.4).

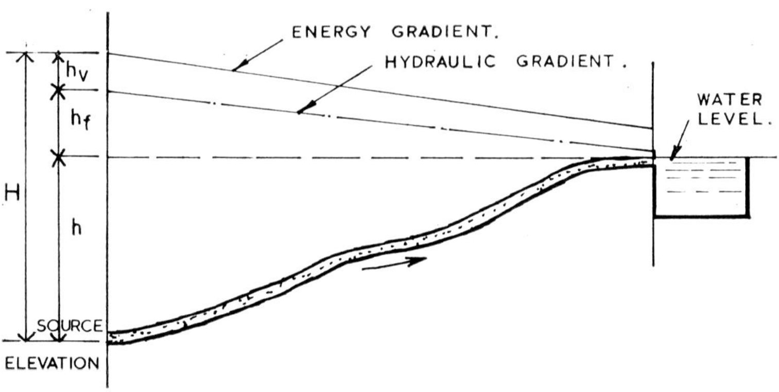

Fig. 1.4 Elevated discharge.

The head conditions are now reversed, and the head contribution H must equal the physical lift plus the losses in the system plus any residual velocity head required, i.e.:

$$H = h + h \ + h_v$$

(iv) *Pressurised discharge systems*

If, in the above example, the fluid were transferred to a vessel under air pressure, an additional head h_p would be needed from the external source (Fig. 1.5).

The jet energy previously attributed to h_v has now to be maintained against this vessel pressure, and the equation becomes:

$$H = h + h_f + h_v + h_p$$

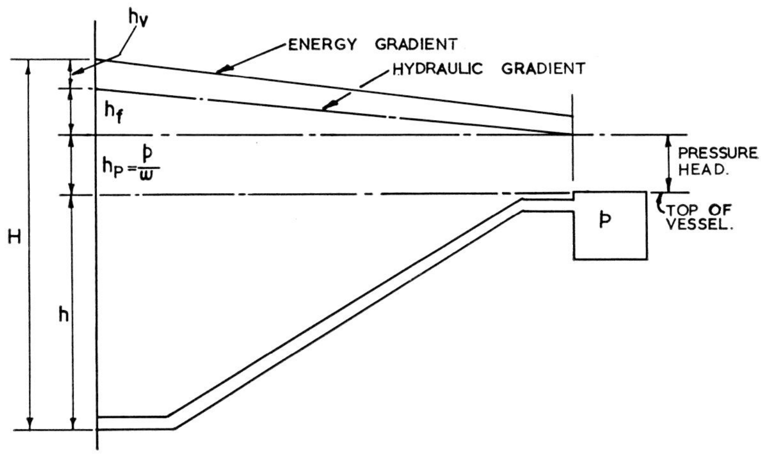

Fig. 1.5 Delivery to pressure vessel.

(v) *Syphonic systems*

A syphon is a system of pipework in which flow is occurring above the hydraulic gradient under the action of gravity (or gravity simulated by a pump) between inlet and outlet.

Fig. 1.6 shows such a gravity system in which atmospheric pressure p_a acts upon a syphon with vertical legs L_1 and L_2.

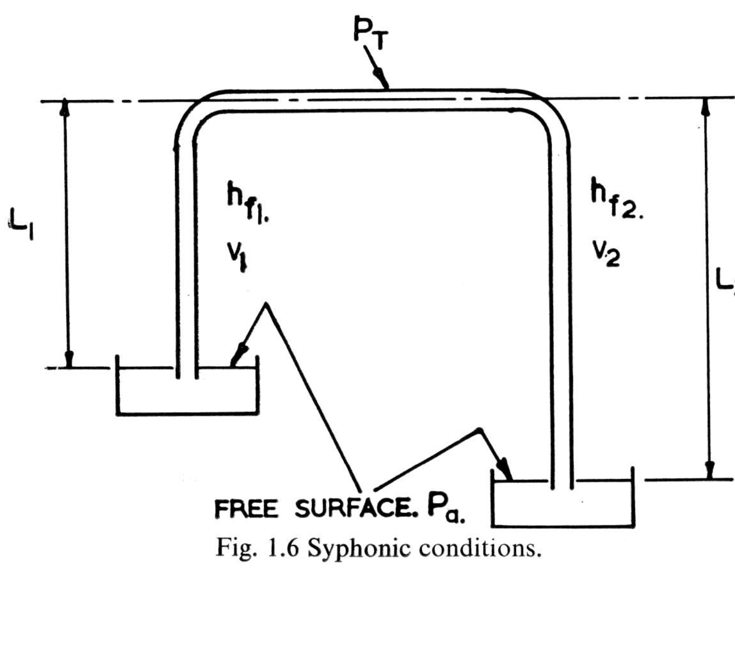

Fig. 1.6 Syphonic conditions.

Consider first the vertical leg L_1 under a steady flow v_1 which causes a head loss h_{f1} over the distance L_1. The column L_1 is in balance with the atmosphere, so the necessary pressure p_T at the top of the syphon is:

$$p_T = p_a - h_{f1} - L_1$$

Each item must be in its appropriate units, so that, if the units are kg/cm² for p_a and m for h_f and L_1, density corrections must be applied as follows:

$$\frac{p_T}{w} = \frac{p_a}{w} - h_{f1} - L_1 - \frac{v_1^2}{2g}$$

where w is the density of the fluid (kg/cm³)
It will be seen that for the maximum lift $p_T/w = O$

$$\frac{p_a}{w} = h_{f1} + L_1 + \frac{v_1^2}{2g}$$

and at no flow

$$\frac{p_a}{w} = L_1$$

With a water density of 1 g/cm³ at normal temperature, and atmospheric pressure at 1·02 kg/cm²

$$L_1 = \frac{1{\cdot}02 \times 1000}{1 \times 100} = 10{\cdot}2 \text{ m}$$

For the outlet leg L_2, however, conditions are slightly different in that the friction drag is of opposite sign, as is the velocity head, so:

$$\frac{p_T}{w} = \frac{p_a}{w} + h_{f2} - L_2 + \frac{v_2^2}{2g}$$

giving identical conditions for zero flow.
Over the whole system we have

$$\frac{p_a}{w} - h_{f1} - L_1 - \frac{v_1^2}{2g} = \frac{p_a}{w} + h_{f2} - L_2 + \frac{v_2^2}{2g}$$

$$\text{so that } (L_2 - L_1) = (h_{f2} + h_{f1}) + \frac{(v_2^2 + v_1^2)}{2g}$$

In other words, the difference in elevation between inlet and outlet is required to overcome friction in the system and velocity generation, as previously.

It should be noted that the above analysis, given in its simplest form, assumes a constant density over the syphon. This is not so, since, as the fluid rises to a region of low pressure, the vapour pressure of the liquid is approached, and there is an increasing tendency for the liquid to gasify. At the top of a syphon there may therefore be a mixture of gas and liquid whose density and characteristics are quite different from those of the parent liquid. The mixture is much lighter, and consequently much higher lifts are in fact possible than, for instance, the 10·2 m often quoted with water. The proper analysis must therefore include for a mean density, to obtain a reasonably true picture of a syphon's behaviour. Data on this aspect is, however, very limited (see Kelly's paper) and safe practice has over the years kept syphonic lifts in various systems to about 8 m.

Research is developing into this important aspect of the behaviour of syphons, and the reader should be vigilant for new data as a result.

The hydraulic gradient relative to a syphon is shown in Fig. 1.7.

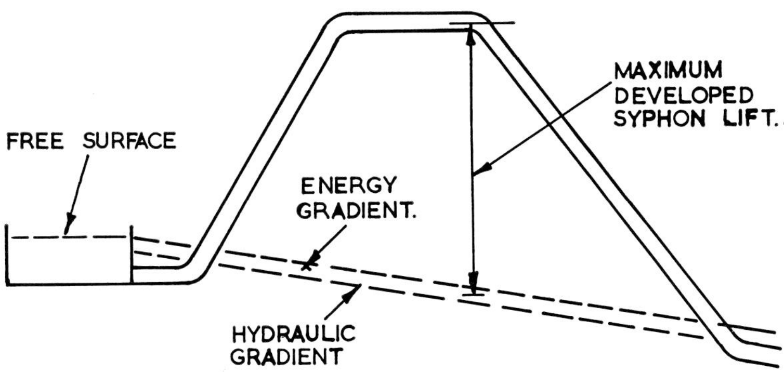

Fig. 1.7 Syphon gradient.

(b) Flow conditions

Conditions of flow fundamentally affect the losses within any system, previously referred to as h_f. These losses occur due to friction between the pipe and the fluid; internal friction within the fluid; and losses induced by disturbance of the smooth flow of fluid at entry, exit, valves, enlargements, bends and by other obstructive influences in the pipeline system.

All such items contribute to a loss of head in the system that is directly proportional to some power of the velocity at which the fluid flows in the pipe (See Fig. 1.8 and subsection (d)). If the rate of flow is plotted against the loss incurred, there is generally a point (regardless of the fluid concerned) below which the loss is linear with respect to the velocity, giving **laminar flow conditions**. Beyond this, the loss increases as a power of the velocity, giving **turbulent flow**.

In order to define these two zones, use is made of a relationship between the absolute viscosity (μ) of a fluid and its velocity of flow (v) in the containing pipework of diameter D. This relationship is in terms of the non-dimensional Reynolds Number (R_e) where

$$R_e = \frac{vDw}{\mu}$$

When using this relationship, take care that the units are correct. Viscosity is defined as the resistance of a liquid in motion to shear stress

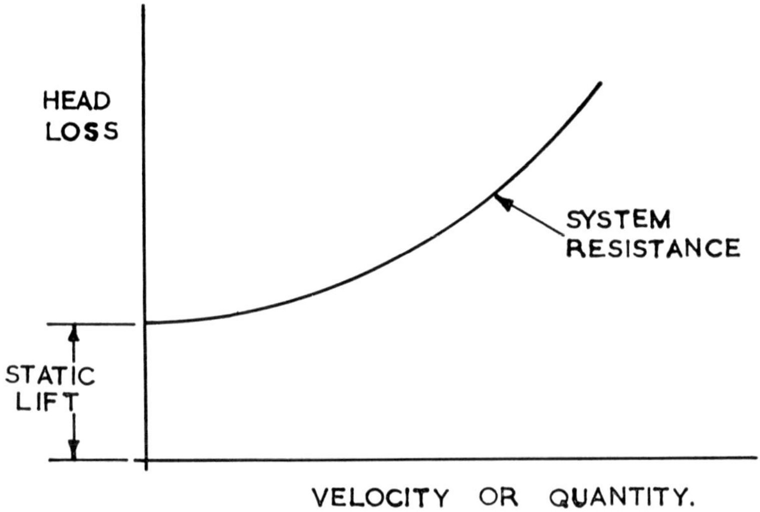

Fig. 1.8 System resistance.

between one layer of the liquid and an adjoining layer parallel to the direction of flow. In metric units, this means the force that a $1/cm^2$ surface moving at 1 cm/s exerts on a parallel surface of similar area at a distance of 1 cm. The viscosity so defined is referred to as the **Absolute or Dynamic Viscosity** and should not be confused with other viscosity definitions, which are, in general, derivatives related to methods of measurement.

Absolute viscosity units then become those of force per unit area for unit time, ie:—

$$\text{force} \times \frac{1}{\text{area}} \times \text{time} = \frac{ML}{T^2} \times \frac{1}{L^2} \times T = \frac{M}{LT}$$

In British units, this means lb./ft.sec., and in metric units g/cm.s, referred to as **poise**. For practical purposes, the poise is too large a unit, hence the use of the **centipoise**, which is a one-hundredth part of a poise. The SI unit adopted is the Ns/m^2 and 1 centipoise equals 10^{-3} Ns/m^2.

It is often desirable to quote the absolute viscosity in terms of a further quantity, and the term **Kinematic Viscosity** (γ) is used to define the absolute viscosity when divided by the weight density (specific weight) of the fluid.

That is,

$$\gamma = \frac{\mu}{w} = \frac{M}{LT} \times \frac{L^3}{M} = \frac{L^2}{T}$$

and

$$R_e = \frac{vD}{\gamma}$$

Again, the units are important and the British units are ft^2/s and the metric units cm^2/s, referred to as a **stoke**. This, for reasons of unit size, is further subdivided into the **centistoke**, a one-hundredth part of a stoke. The SI unit is m^2/s and 1 centistoke $= 10^{-6}$ m^2/s.

Since the Reynolds Number is a product of

$$\frac{L}{T} \times L \times \frac{T}{L^2}$$

it is dimensionless, and the value of R_e in consistent units will be the same.

Reynolds found that at values of R_e of 2,000 and below, flow was laminar; at 3,000 and above, turbulent. In nearly all normal circumstances with the more usual fluids, flow is turbulent, as can be seen in the following calculation.

Example: Water flows through a 150 mm pipe and the kinematic viscosity is $1{\cdot}21 \times 10^{-6}$ m^2/s. The equivalent limit of laminar flow velocity is therefore:

$$v = \frac{R_e \gamma}{D} = 2000 \times \frac{1000}{150} \times 1{\cdot}21 \times 10^{-6} = 0{\cdot}016 \text{ m/sec}$$

Since normal velocities of flow in pipework are around 1 to 3m/s, it is clear that normal flow is turbulent.

Conversion from kinematic viscosity to absolute viscosity can be made from Fig. 1.9, and conversion of different unit systems from Figs. 1.10, 1.11 and 1.12. The viscosities of common liquids are given in Figs. 1.13 and 1.14.

(c) Pipeline losses

Losses due to flow through the pipe system are entirely parasitic and fall into the following categories:

(i) *Pipe friction*

There are many formulae, designed to give accurate expressions for head loss in certain defined conditions in various types of confining pipework. Some are elaborate, and the reader should be wary of relying too much on formulae of a 'laboratory' nature which, when applied to a practical pipework system, may give disappointing results. It is essential to realise that hydraulics is an inexact science, in so far as pipe losses are concerned, and although every care must be taken that a pipe system is analysed properly for friction and other losses, practical considerations must also be borne in mind. It is of little satisfaction to the operator to find minor modifications (which may be needed in practice) beyond the capacity of the system. Nor is it of much satisfaction to a designer to analyse a system very minutely in ignorance of its operating conditions. Between operation and design there is in most systems a good margin of flexibility, which must always be borne in mind if a good, sensible, practical design is to result. Within these limits, certain practical formulae have found general acceptance, from which charts of pipe sizing and losses have been developed. Although the most 'accurate'

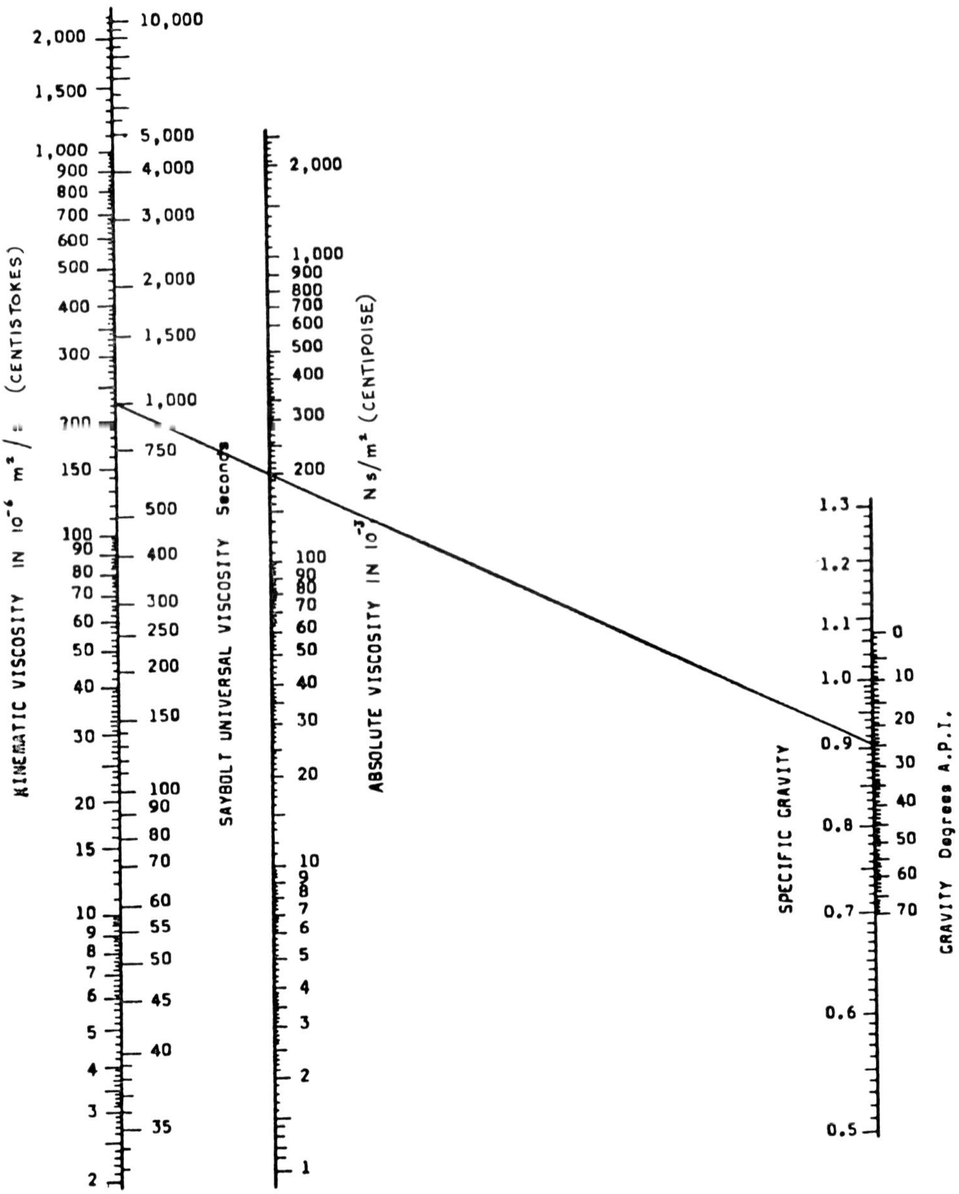

Fig. 1.9 Viscosity conversion.

Example using Fig. 1.9: A fluid of absolute viscosity 200 centipoise and specific gravity 0·9 will have a kinematic viscosity of 225 centistokes or 1000 Saybolt seconds.

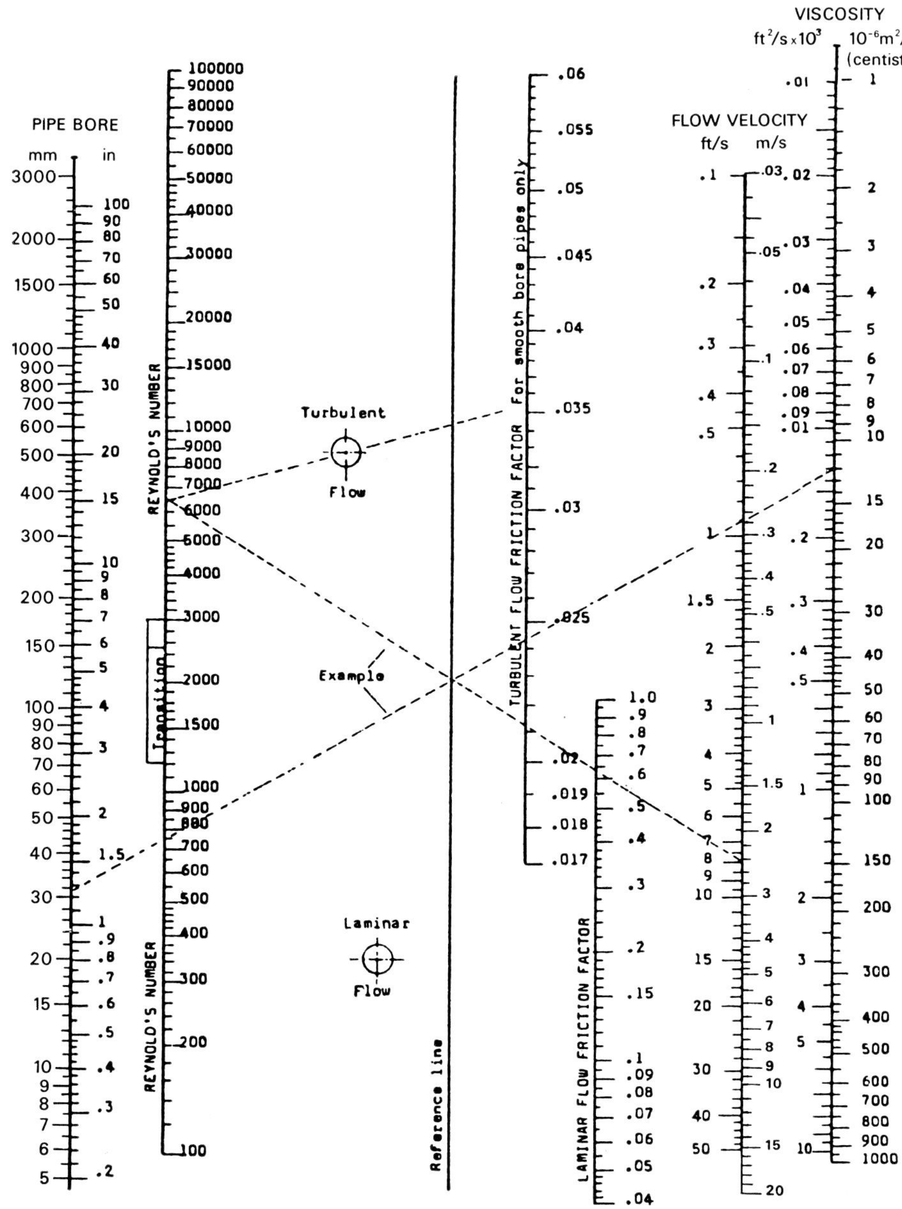

Fig. 1.10 Reynolds number and friction factor.

Example using Fig. 1.10: A fluid of viscosity 12 centistokes flows in a pipe 31 mm bore at 2·5 m/s. Find the friction factor f. Connect the 12 centistokes point to the 32 mm bore and note the intercept on the reference line. Produce the line from the 2·5 m/s point through the intercept to read 6500 Reynolds Number—clearly turbulent flow. Connect this point through the guide point for turbulent flow to give f = ·0355.

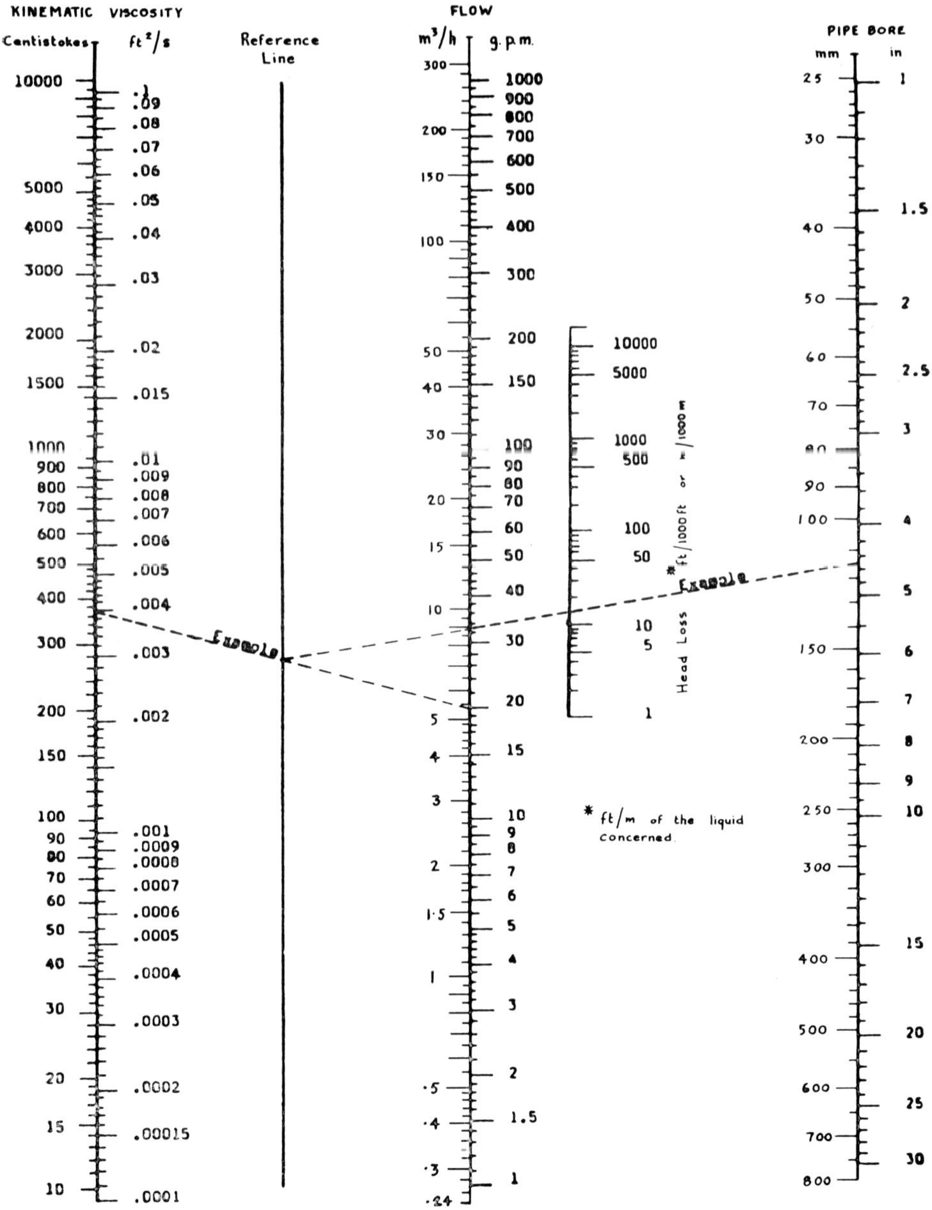

Fig. 1.11 Head loss with laminar flow.
(Not suitable for water)

Example using Fig. 1.11: A fluid of viscosity 370 centistokes flows at 5·5 m³/h in a 114 mm pipe. Find the head loss. Join the 370 centistokes point to the 5·5 m³/h point and note the intercept on the reference line. Join this point to the 114 mm bore point, giving a head loss of ·015 of fluid m per m run of pipe.

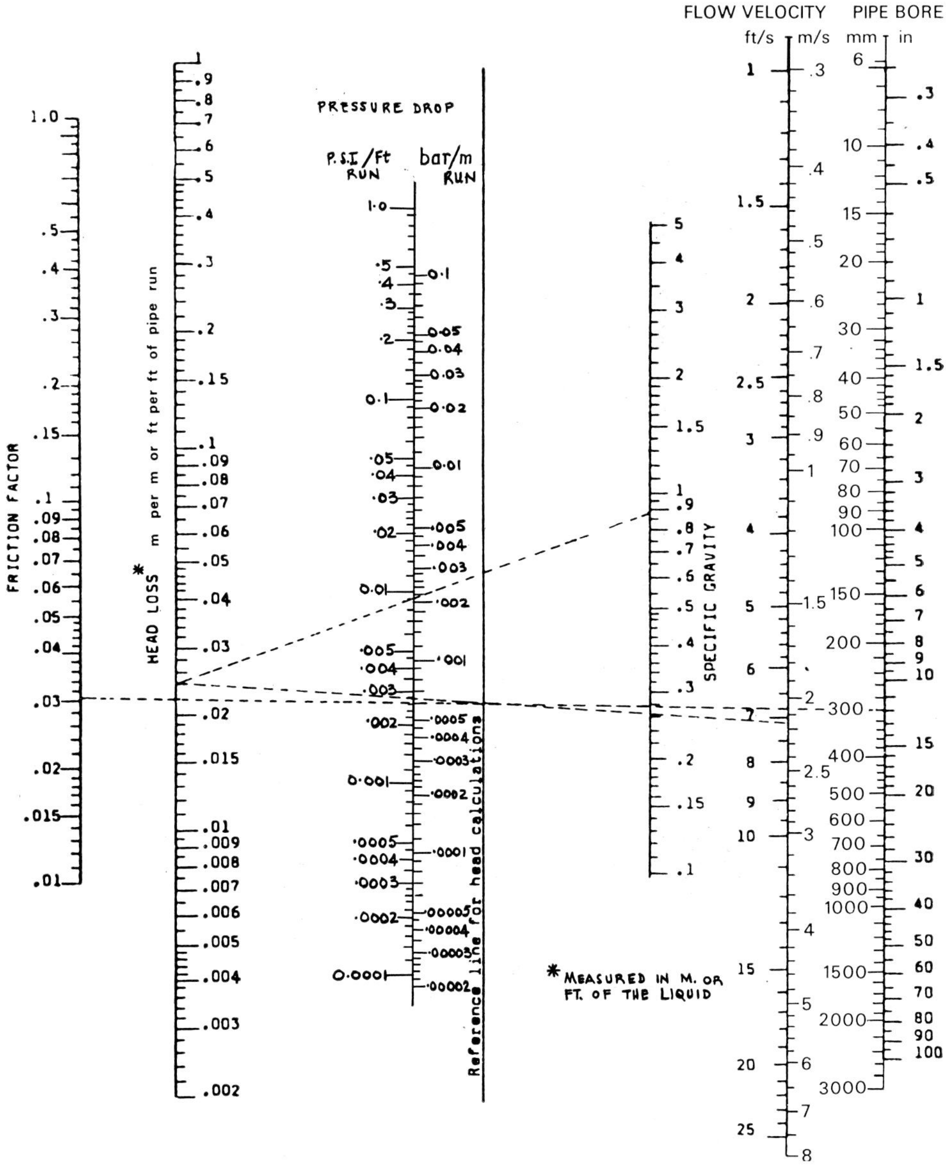

Fig. 1.12 Head loss and pressure drop with turbulent flow.

Example using Fig. 1.12: A fluid S.G. 0·9 flows at 2·15 m/s in a 300 mm pipe, friction factor ·031. Find the pressure drop or head loss. Join the friction factor to the pipe bore and note the reference intercept. Through this point produce a line from the 2·15 m/s point to cut the Head Loss scale at a value 0·024 m per m run of pipe. From this point join to the 0·9 S.G. point and note the head loss in bar per m run of pipe (0·0021).

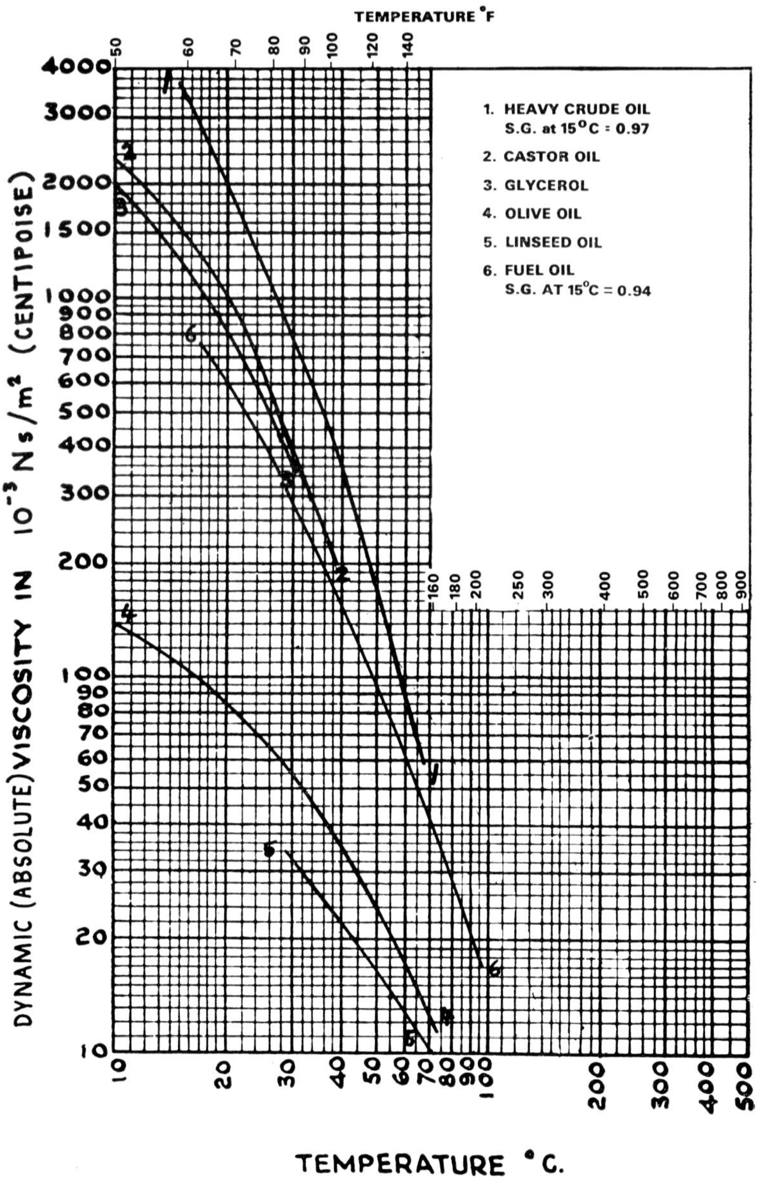

Fig. 1.13 Temperature-viscosity curves for high-viscosity liquids.

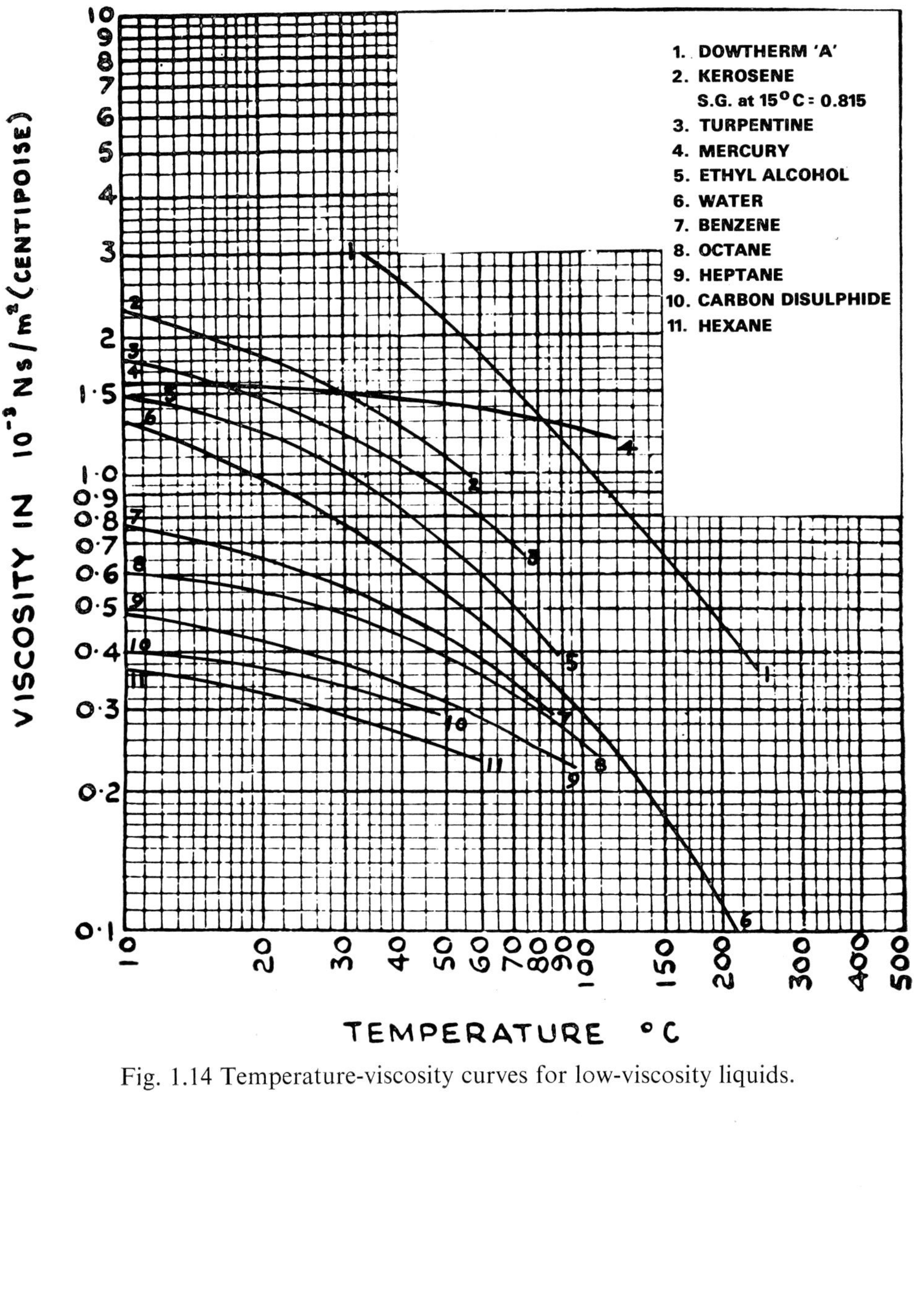

Fig. 1.14 Temperature-viscosity curves for low-viscosity liquids.

formula may well be that devised by Colebrook and White, its presentation in engineering form has posed some difficulty, because its main parameters are not those readily available to the engineer. An exhaustive treatise on the development of friction formulae may be seen in Ackers' work, 'The Resistance of Fluids flowing in channels and pipes' (HMSO – Paper No. 1, D.S.I.R., 1958), in which he develops suitable parameters and graphs for the solution of the Colebrook–White equation on a practical basis giving a satisfactory engineering presentation.

Perhaps the best known fundamental equation covering pipe losses is the Manning or Darcy formula

$$h_f = \frac{fLv^2}{2gD}$$

where f = friction factor (dimensionless)
L = length of pipe considered (m or ft)
D = pipe diameter (m or ft)
g = gravitational acceleration (m/s^2 or ft/s^2)

This formula has the advantage that it may be used for all liquids under all conditions of flow. Its acceptability in different conditions depends on a satisfactory value of f being available in each case.

Where laminar flow applies, the factor is given by the expression

$$f = \frac{64}{R_e}$$

and the Reynolds Number (R_e) itself varies with the velocity of flow, the pipe diameter and the kinematic viscosity.

For turbulent flow conditions, the value of f depends on the relative roughness of the pipe wall with respect to the diameter of the pipe. This function is related to the Reynolds Number for a series of values for friction coefficients based on the Colebrook–White equation covering the transition zone of turbulent flow, such that:

$$\frac{1}{\sqrt{f}} = -2 \log \left[\frac{k}{3{\cdot}7D} + \frac{2{\cdot}51}{R_e\sqrt{f}} \right]$$

The solution is presented in Fig. 1.15. Roughness factors k, based on experimental results (see Table 1.1), are allocated to each type of pipe surface and the relative roughness $\frac{k}{D}$ established. The Reynolds Number is calculated and in conjunction with k/D enables the friction factor f to be read from Fig. 1.15.

Finally f is again used in the expression

$$h_f = \frac{fLv^2}{2gD}$$

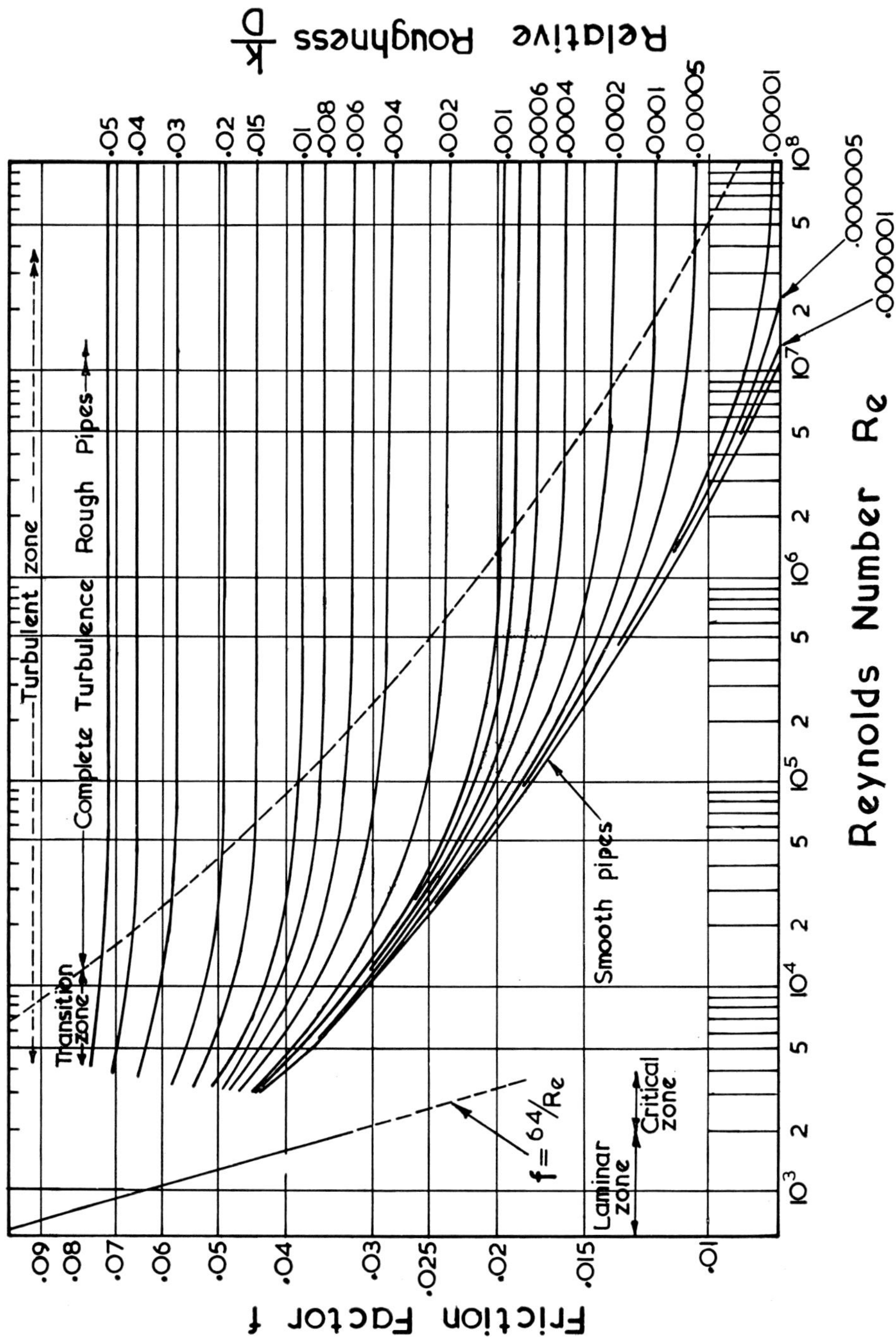

Fig. 1.15 Friction factor f for turbulent flow. Value of f as in $hf = \frac{fLv^2}{2gD}$

to obtain the head loss in the system.

Care must be taken that the values of f used are in fact related to the formula in its above form. Some sources use $h_f = \frac{4fLv^2}{2gD}$ and f factors for such a version would of course be $\frac{1}{4}$ of those used here. It is also necessary to note that k and D must be in the same units, since $\frac{k}{D}$ must be dimensionless.

Table 1.1 gives values of k for use with Fig. 1.15. These are values for pipes in clean conditions.

TABLE 1.1. HYDRAULIC ROUGHNESS

(1) Class of surface	(2) k value for use with Fig. 1.15		(3) Multiplying factor, Fig. 1.10
	m	ft	
Drawn, non-ferrous plastics	Smooth to 0·000003	Smooth to 0·00001	1·0
Pipes with spun bitumen lining	0·00003	0·0001	1·1
Pipes with spun concrete lining	0·00003	0·0001	
New drawn steel	0·00003	0·0001	
Coated (dipped) steel	0·00006	0·0002	1·2
Coated (dipped) cast iron	0·00016	0·0005	1·5
Galvanised iron	0·00016	0·0005	
Uncoated cast iron	0·00033	0·001	1·4
Glazed stoneware	0·00066	0·002	1·8
Precast concrete	0·00066	0·002	
Rough concrete with poor joints	0·0033	0·01	3·0
Old pipes, very rough	0·0066–0·016	0·02–0·05	4–5

Within a more limited application range, Fig. 1.10 enables the 'smooth pipe' friction factor to be obtained directly. Variations from the 'smooth case' friction factor are difficult to assess since the multiplying factor to be applied varies with R_e and pipe relative roughness.

However, for R_e values of the order of 10^5 and pipe sizes of about 152 mm., the multiplying factors quoted in Table 1.1 opposite the of class pipe concerned may be used.

Judgement is also necessary, of course, in the use of Fig. 1.15 and Figs. 1.9–1.12. If the pipe concerned is likely to deteriorate quickly in service, then values of k higher than those applicable to the new pipe must be used in calculating the head loss.

A formula widely used in the water industry, and very convenient in graphical form, is that due to Hazen-Williams which gives:

$$v = 0.3351\ C\ (D)^{0\cdot 63}\ (h_f/L)^{0\cdot 54}$$

where C = a pipe friction constant, dependent on the type, condition and diameter of the pipe
v = the velocity of flow (m/s)
D = the internal diameter of the pipe (m)
h_f = the head loss due to friction (m)
L = the pipe length (m)

This formula, presented in the universal pipe friction diagram of Fig. 1.16, will be found useful in the direct determination of velocities in various pipe diameters, as well as friction losses. Note that the value C of this formula corresponds to adjustments made to the value of f in the previous formula, to cover varying pipe conditions, in which **water** is flowing at normal temperatures.

(ii) *Losses in pipe fittings*

So far, only friction losses in the pipe itself have been considered. In addition to these, allowances have to be made for equivalent losses at valves, bends, tees, etc., which inevitably introduce themselves into any practical pipe assembly, together with losses that occur at the system inlet and exit. The assessment of loss in such fittings can be accomplished either by analysing the loss and adding it on to the h_f previously calculated, or, much more conveniently, by determining an 'equivalent' pipe length. In this case, each fitting is considered as an added length of pipe, and the total 'equivalent' pipe length then used in the h_f calcula-

TABLE 1.2. VALUES OF $v^2/2g$

(1) Flow velocity	(2) $v^2/2g$	(3) Velocity	(4) $v^2/2g$
ft/sec	ft	ft/sec	ft
1·0	0·015	7·0	0·76
2·0	0·062	8·0	0·99
3·0	0·14	9·0	1·26
4·0	0·25	10·0	1·55
5·0	0·39	11·0	1·88
6·0	0·56	12·0	2·24
m/s	m	m/s	m
0·25	0·003	2·00	0·204
0·50	0·011	2·25	0·256
0·75	0·029	2·50	0·318
1·00	0·051	2·75	0·385
1·25	0·079	3·00	0·459
1·50	0·115	3·25	0·523
1·75	0·156	3·50	0·622

IMPERIAL $V = 0.5512\ C\ (D)^{0.63}(H/L)^{0.54}$

METRIC $V = 0.3351\ C\ (D)^{0.63}(H/L)^{0.54}$

V = VELOCITY FT/S D = INT. DIA FT H/L SAME UNITS

V = VELOCITY M/S D = INT. DIA. M H/L SAME UNITS

C IS A PIPE FRICTION FACTOR OF VALUE DEPENDENT ON PIPE, COMMON IN VALUE FOR METRIC AND IMPERIAL UNITS

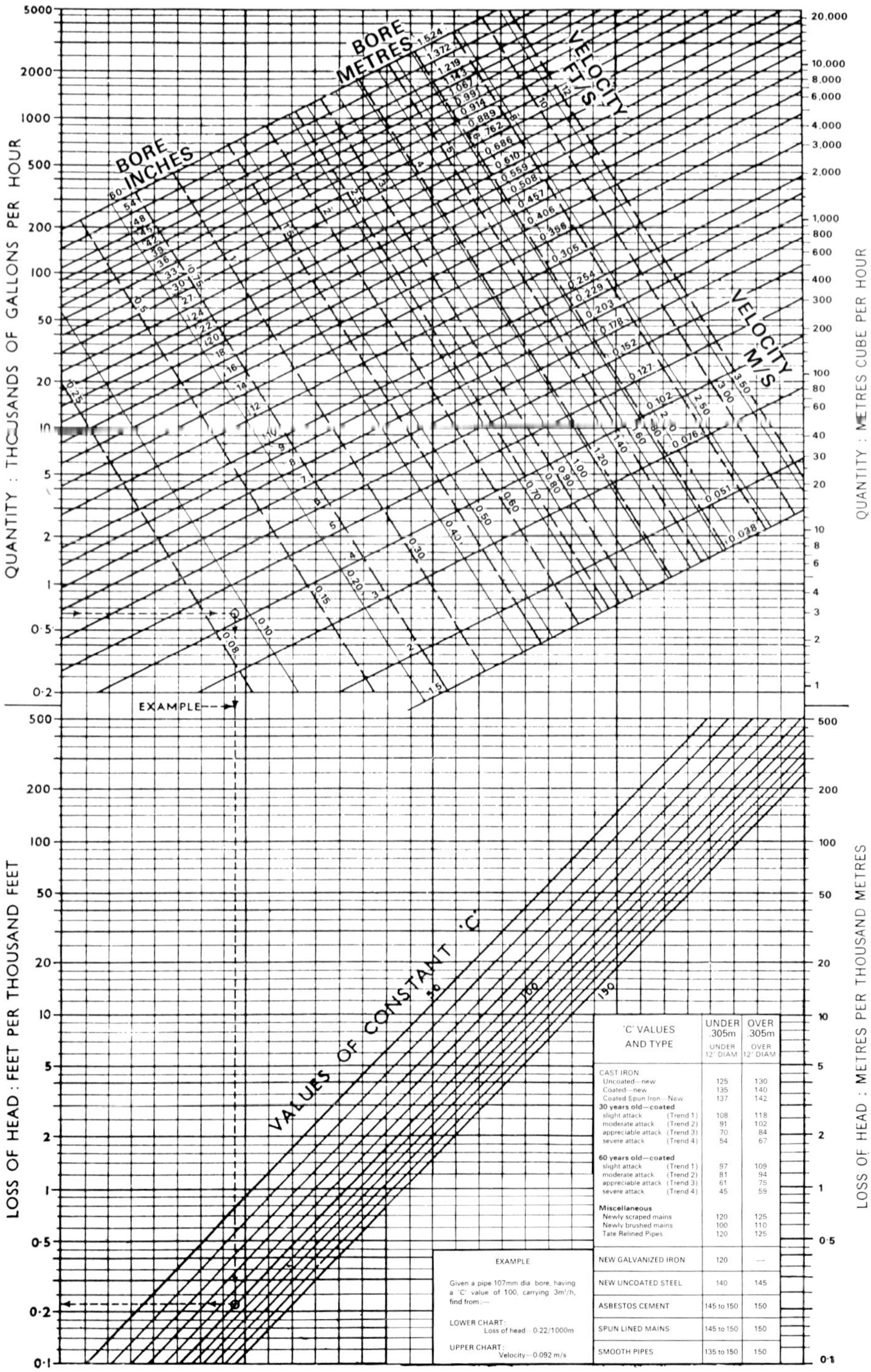

'C' VALUES AND TYPE	UNDER .305m UNDER 12" DIAM	OVER .305m OVER 12" DIAM
CAST IRON		
Uncoated—new	125	130
Coated—new	135	140
Coated Spun Iron—New	137	142
30 years old—coated		
slight attack (Trend 1)	108	118
moderate attack (Trend 2)	91	102
appreciable attack (Trend 3)	70	84
severe attack (Trend 4)	54	67
60 years old—coated		
slight attack (Trend 1)	97	109
moderate attack (Trend 2)	81	94
appreciable attack (Trend 3)	61	75
severe attack (Trend 4)	45	59
Miscellaneous		
Newly scraped mains	120	125
Newly brushed mains	100	110
Tate Relined Pipes	120	125
NEW GALVANIZED IRON	120	—
NEW UNCOATED STEEL	140	145
ASBESTOS CEMENT	145 to 150	150
SPUN LINED MAINS	145 to 150	150
SMOOTH PIPES	135 to 150	150

EXAMPLE

Given a pipe 107mm dia. bore, having a 'C' value of 100, carrying $3 m^3/h$, find from:—

LOWER CHART:
Loss of head—0·22/1000m

UPPER CHART:
Velocity—0·092 m/s

Fig. 1.16 Universal pipe friction diagram for use with water.

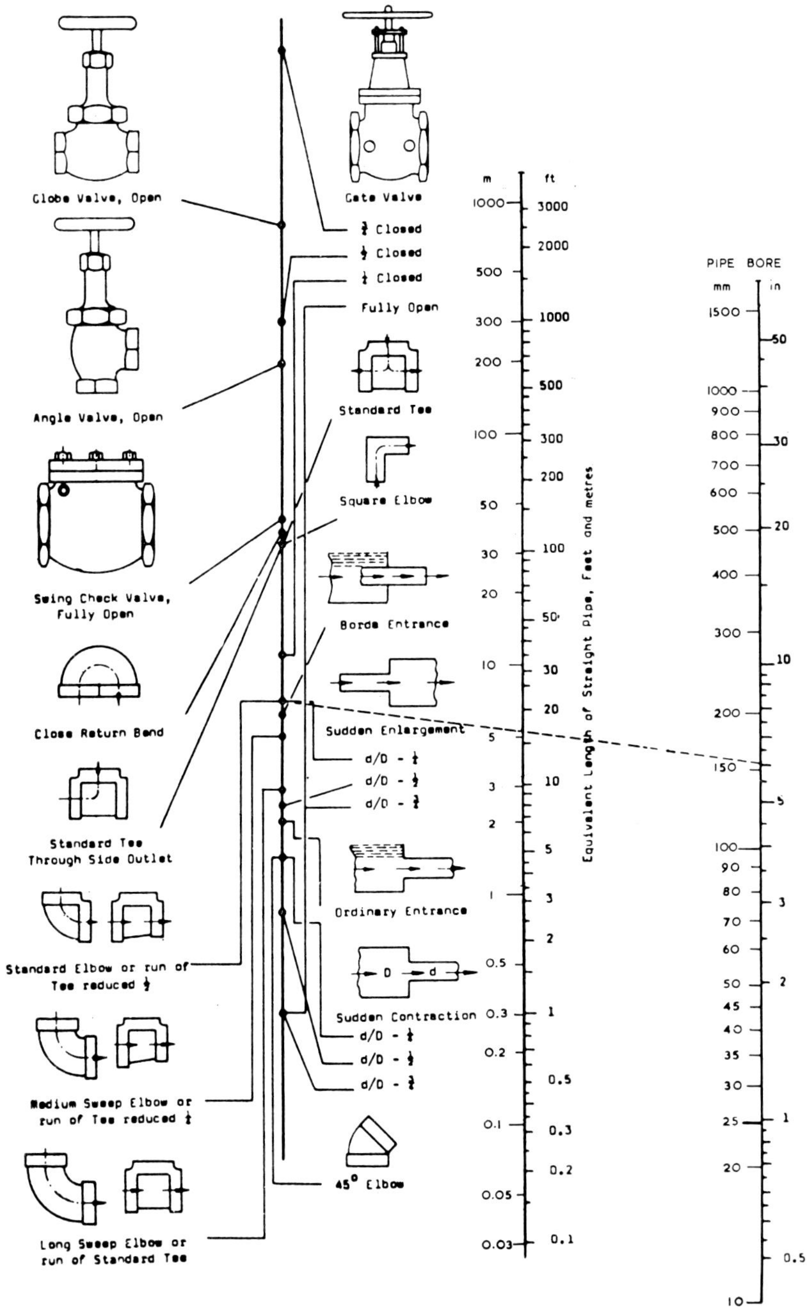

Fig. 1.17 Resistance of valves and fittings.
(See Section Five (c) for a worked example)

tion. Table 1.1 and Fig. 1.17 give such accepted practical 'equivalent' values, either as a multiplying factor on pipe diameter or a direct reading of equivalent length (m.)

(iii) *Velocity generation*

Table 1.2 provides a useful direct reading for the head allowance to be made for velocity on the basis

$$h_v = \frac{v^2}{2g}$$

and it will be noted that since this also is a function of v^2, it can be considered jointly with pipe friction losses. When once these integrated losses have been calculated for a particular velocity, their variation for any other velocity is parabolic.

(iv) *Effects of age or deterioration*

Much has been written on the ageing of pipes in service. Corrections to roughness factors based on complicated functions of water chemistry, rate of growth of slime, etc., have been devised, but little of practical use has developed from such exercises, except in specific cases of pipe materials used in certain areas. Most organisations have developed their own margins to cover ageing factors, but even with good evidence there is always the economic consideration whether it is better to make only a small allowance and economise in pumping costs, or whether full allowances should be made at the start and any pumping capacity throttled until it is required. With the increasing use of inhibitors and good control of fluid quality, now in general practice, it is not, broadly speaking, economic to put heavy penalties on pumping head to anticipate pipework ageing. It is better to ensure that there is adequate possibility of modifying the pump/motor combination in due course, and not to accept pump designs that are pared down to the minimum.

The above comments, coupled with the established practical methods of pipe loss calculations, serve to emphasise the general sensible tolerances to be built into a design of pipe system. The reader should, on the other hand, realise that specialised problems of hydraulics may well benefit from established formulae designed for a particular purpose – water supply through rock tunnels, for example – and such problems should always be tackled on the basis of their speciality.

(d) System resistance characteristics

From the above, it will be clear that the system is subjected to static head, pressure, and head losses dependent on velocity, together with any residual heads required for operational purposes. These are graphically presented for any system in Fig. 1.8, referred to briefly earlier.

As the velocity increases, h_f losses increase parabolically and the curve resulting as a plot of head against velocity or quantity is called the 'system resistance'. At any value of v (or Q) it shows the head contribution which is necessary to the system if flow is to be maintained at that particular velocity or quantity.

Usually h_v is very small compared with h_f and, although strictly h_v has to be included, it is often neglected. The value h_f and h_v may frequently be referred to as h_f only, for simplicity in diagrams.

SECTION TWO

THE PUMP

(a) Performance of basic pump types

Rotodynamic pumps fall into three main classes, based on the path the water takes through the pump. These are:

(i) *Axial flow pump*

As its name suggests, the axial flow pump propels the water through the pump casing with the minimum of angular deviation. The pump blades are of the propeller type, and both entry and discharge from the propeller are in-line with one another. Although, for reasons of pump construction, the final discharge path may be at an angle to the entry (as illustrated in Fig. 2.1), essentially the movement of the water is axial,

DUTY	MAIN CIRC.
Capacity	4600 m^3/h
Total head	4 m
Speed range	500/700 rev/min
Liquid pumped	Sea water
Motor rating	100 kw

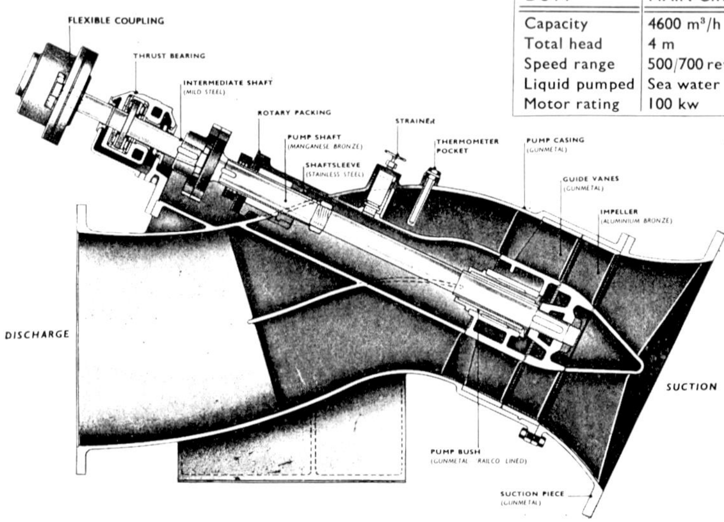

Fig. 2.1 Axial pump.

and the energy is imparted to the water by virtue of the blade shape. These pumps are designed to move large quantities through a small increase in head.

(ii) *Centrifugal pump*

At the other extreme, that of the centrifugal or radial pump, the energy is transferred by a centrifugal thrust applied to the water in the rotor or impeller. Water enters into the eye of the impeller and is then guided in the passage between the stationary casing and the rotating impeller in a radial direction. The thrust is given to the water in a series of vanes on the impeller, the shapes of which are designed to give an efficient increase in velocity as the water moves towards the periphery and is finally discharged into the volute area of the casing. A diffuser may be incorporated to provide for the increase of head in certain pumps of this kind (see Fig. 2.2 (a) and (b)). The radial flow pump is adaptable to most duties, but is essential where high heads are concerned.

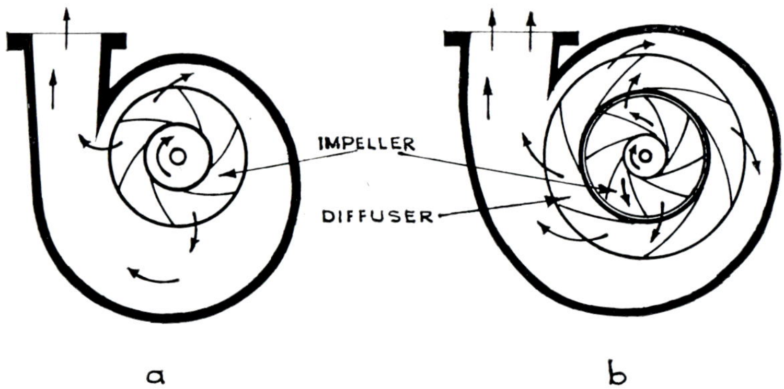

Fig. 2.2 Radial flow pump: (a) with volute; (b) with diffuser vanes and volute.

(iii) *Mixed flow pump*

In between the low head limitations of the axial flow pump, and the best use of the radial flow pump at high heads, is to be found an economical combination of the two in the 'mixed flow' type. It may be used with a divergent cone as in Fig. 2.3 (a) or with a volute as in Fig. 2.3 (b).

Each of the three pumps shows a characteristic behaviour in the way it contributes energy to the liquid and absorbs power. A series of such characteristic curves showing the variation from type to type, based on the

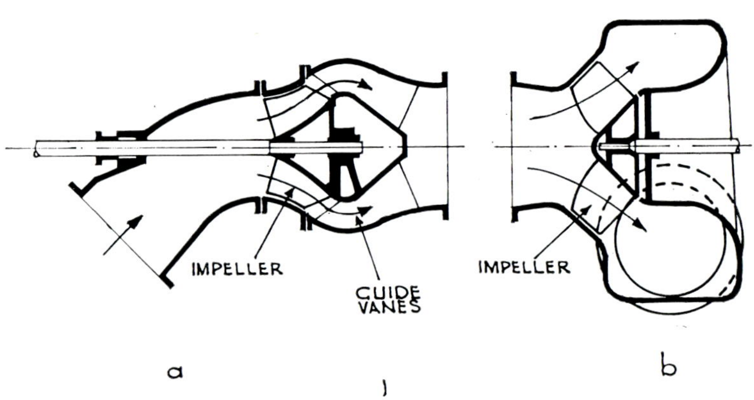

Fig. 2.3 Mixed flow pump: (a) with divergent cone; (b) with volute.

quantity of water passing through the pump, is shown in Fig. 2.4 (a), (b) and (c).

(b) Speed, efficiency and power

(i) *General relationship*

The basic characteristics shown in Fig. 2.4 are for a particular speed. Duty points would ideally be chosen at the vertical ordinate through the Q-H curve at the point of maximum efficiency. Changes in speed produce a near-parallel shift of the Q-H curve, as shown in Fig. 2.5, which also shows points of equal efficiency at each speed change. It will be seen that the points of maximum efficiency form a parabola (dotted) through the origin. If, then, a pump is to be operated at reduced speeds, it is best to reduce also the duty point in both head and quantity.

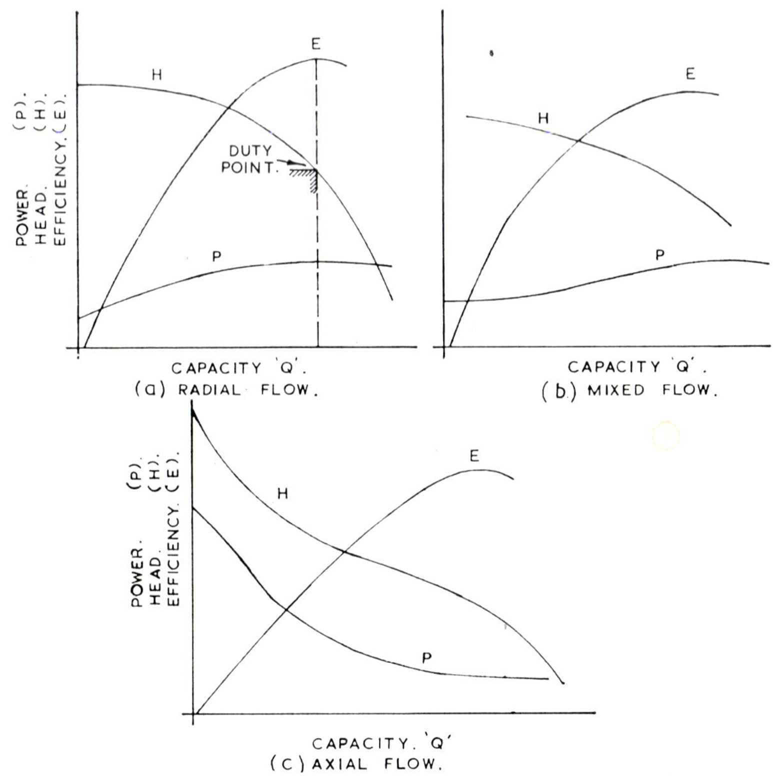

Fig. 2.4 Characteristic curves for various pump types.

H and *Q* changes reflect pump speed *(N)* changes in the following relationships:

$$\frac{Q_1}{Q_2} = \frac{N_1}{N_2}$$

$$\frac{H_1}{H_2} = \left[\frac{N_1}{N_2}\right]^2$$

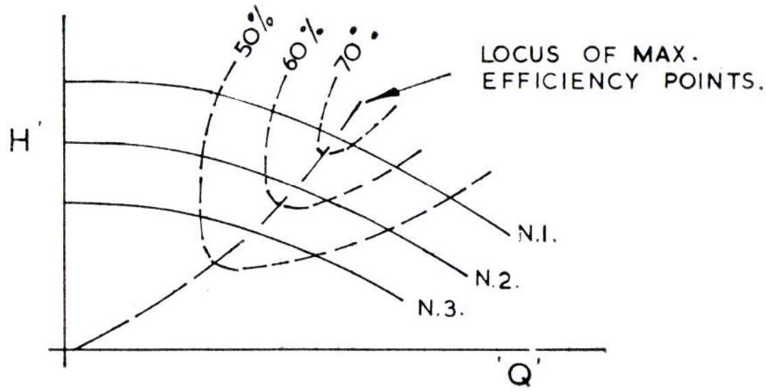

Fig. 2.5 Effects of speed variations

It follows therefore, since pump power requirement *(P)* is a direct function of Q and H, that:

$$\frac{P_1}{P_2} = \left[\frac{N_1}{N_2}\right]^3$$

Duty points can also be moved by changes in impeller diameter *(d)*. Such a change effectively means that the angular velocities at the impeller vane-tip changes directly with $\frac{d}{2}$. The change of diameter is therefore similar in effect to a direct change of speed and affects Q and H in the same manner.

$$\frac{Q_1}{Q_2} = \frac{d_1}{d_2}$$

$$\frac{H_1}{H_2} = \left[\frac{d_1}{d_2}\right]^2$$

and again

$$\frac{P_1}{P_2} = \left[\frac{d_1}{d_2}\right]^3$$

From the above properties, pump performances at low speed (where test conditions are limited in both power and quantity) can be used to forecast the performance of the same pump in installed conditions.

(ii) *Specific speed*

It should be noted that pumps are often classified into distinctive types by their 'specific speed', regardless of the size of the pump.

In British Units, specific speed n_s is related to a discharge of one gal/min. through 1 ft head, so that:

$$n_s = \frac{NQ^{\frac{1}{2}}}{H^{\frac{3}{4}}}$$

the units being consistent, i.e.: N, n_s (r.p.m.), Q (gal./min.) and H (ft.). In SI units, of rpm, m³/h and m, the British values need multiplying by:

$$\frac{(0{\cdot}273)^{\frac{1}{2}}}{(0{\cdot}305)^{\frac{3}{4}}}$$

that is, 1·273, as given in Table 2.2.

TABLE 2.1. SPECIFIC SPEED AND PUMP TYPE

(1) Category	(2) Nature of duty	(3) Specific speed (in rpm, m^3/h and m units)
Centrifugal high lift	Low quantities	1150
Centrifugal low lift	Large quantities	1650–2800
Mixed flow types:		
(a) Francis Impeller	Intermediate duties	3800–5000
(b) Mixed flow impeller		5000–7600
Axial flow types	Low head, high quantities	7600–16500

Fig. 2.6 shows impeller physical form in relation to specific speed, whilst Table 2.2 gives conversion factors from one set of units to another. The Shape Number quoted is dimensionless, as long as consistent units are used, i.e.: rev/sec., m/s^2, m, m^3/s, etc.

When using specific speeds with multiple pumps, take care that, in the case of multi-stage pumps, the head used is the 'head per stage', i.e. $H/3$ in the case of a 3 H stage unit.

(iii) *Power requirements*

The power requirement of the fluid pumped is given by

$$9{\cdot}807\ Qw_1H \text{ watts}$$

with Q (m^3/s), w_1 (kg/m^3) and H (m).

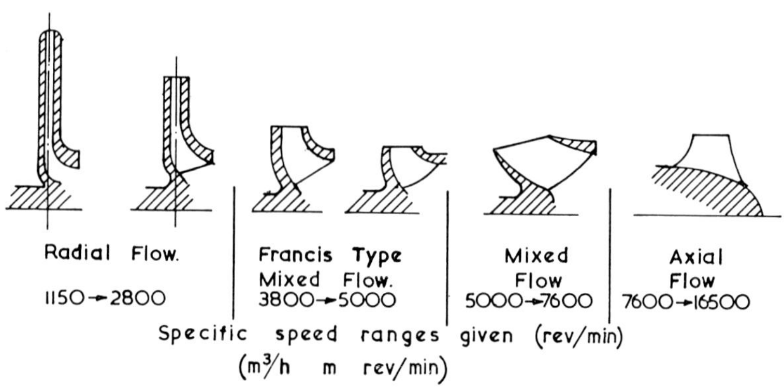

Fig. 2.6 Relationship between radial, mixed, and axial flow impellers in terms of specific speeds.

Efficiency factors have to be added covering pump, couplings and driver unit, so the total input power to the driver unit is given by:

$$\frac{9{\cdot}807\ Qw_1H}{E} \text{ watts}$$

where E is the overall efficiency, expressed as a fraction.

The above, however, is a consideration of power requirement at any duty point. It will be seen from Fig. 2.4 that the shape of the power *(P)* curve is quite different in each type of pump. For a centrifugal pump, the power increases as the quantity pumped; for the axial flow pump the converse is the case, and for mixed flow a variety of power curves is possible, depending on whether the mixed flow pump is predominantly axial flow or centrifugal.

TABLE 2.2. SPECIFIC SPEED VALUES

Units of quantity and Head		(1) gal./min. ft. r.p.m.	(2) ft³/s ft. r.p.m.	(3) U.S. gal./min. ft. r.p.m.	(4) litre/s m r.p.m.	(5) m³/s m r.p.m.	(6) m³/h m r.p.m.	(7) Addison's Shape Number*
Conversion factors relative to unity	(1)	1·0	0·0517	1·096	0·671	0·0212	1·273	0·0637
	(2)	19·34	1·0	21·19	12·98	0·410	24·63	1·232
	(3)	0·913	0·0472	1·0	0·612	0·0194	1·162	0·0581
	(4)	1·490	0·0771	1·633	1·0	0·0316	1·898	0·0949
	(5)	47·13	2·437	51·64	31·62	1·0	60·0	3·0
	(6)	0·786	0·0406	0·861	0·527	0·0167	1·0	0·050
	(7)	15·70	0·813	17·21	10·54	0·333	20·0	1·0

* Shape Numbers are $= \dfrac{100NQ^{\frac{1}{2}}}{(gH)^{\frac{3}{4}}}$

and compatible units would be $Q=\text{m}^3/\text{s}$, $n=\text{rev/sec}$, $g=\text{m/sec}^2$, $H=\text{m}$

When a centrifugal pump is brought on line against a closed valve condition, the load on the driver unit increases to the duty point as the valve is opened. If thereafter the curve becomes convex with very slight or no increase in power demand, it is said to be a non-overloading power curve. In the case of the axial flow pump, if it is started against closed valve conditions, it can be seen from Fig. 2.4 that the power requirement at starting may well be far in excess of the power requirement at duty point – for which the driver unit is usually sized. It is important, therefore, that axial flow pumps are not brought 'on line' against closed valve conditions, and that either a by-pass arrangement be devised or pressure actuated valves installed, to ensure that the pump cannot generate pressures greatly in excess of the duty point.

Power curves for mixed flow pumps are usually quite flat, and from this point of view mixed flow pumps are very easily operated, both singly and in combination.

The matching of the required power curves to a prime mover is discussed in Section Six.

(c) Suction conditions

The above characteristics of a pump, particularly with respect to head-flow, are valid only if the suction conditions to which the pump is subjected are entirely within the pump's capability. In Fig. 2.7 Scheme (1), the pump

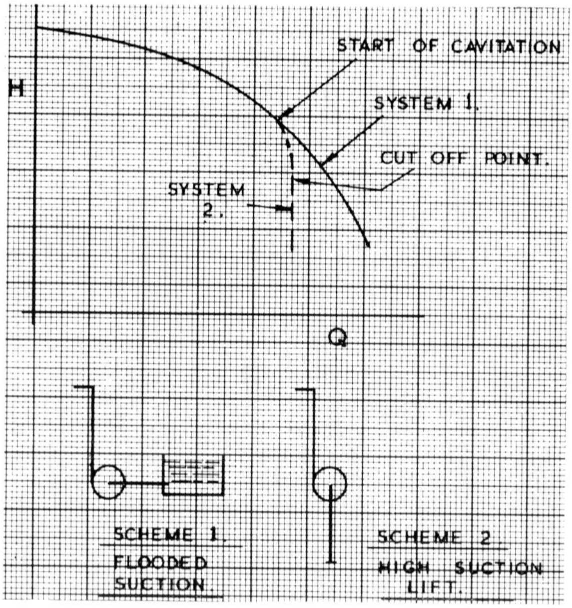

Fig. 2.7 Effects of cavitation on *Q-H* curve.

is operating with a flooded suction and, if h_s is sufficient, with normal velocities no interference will be seen with the continuity of the head capacity curve.

If, however, the pump is subjected to a large suction lift, either of pure elevation or elevation combined with heavy friction losses, such as is shown in Scheme 2, considerable interference will be seen with respect to the *Q-H* curves.

This interference, which takes the form of a sharp cut-off as indicated, occurs when the pressure in the liquid entering the 'eye' of the impeller falls below its vapour pressure at the temperature concerned, and cavitation occurs. The liquid then vaporises and the effect on the pump is that, instead of a homogeneous liquid passing through the impeller, air and vapour bubbles separate out in the liquid which then becomes a mixture of vapour, air and liquid. At the cut-off point, the amount of vapour/air has become such that the pump cannot increase its capacity in terms of liquid, and cavitation is complete. It is important to realise that what matters to the pump impeller is this continuity of fluid. Provided it remains so, the fact that the fluid is at sub-atmospheric pressure is not in itself of any significance. It will be appreciated that atmospheric pressure being say 1000 mbar at sea level, the equivalent head of water will be about 10·2m, and hence, neglecting friction, a pump is theoretically capable of lifting water through this vertical distance before the pressure is so reduced at the eye of the impeller that the water starts to vaporise.

In practice, however, losses occur which reduce the 10·2m: friction losses in a suction pipe, velocity generation and impeller eye losses. To ensure that

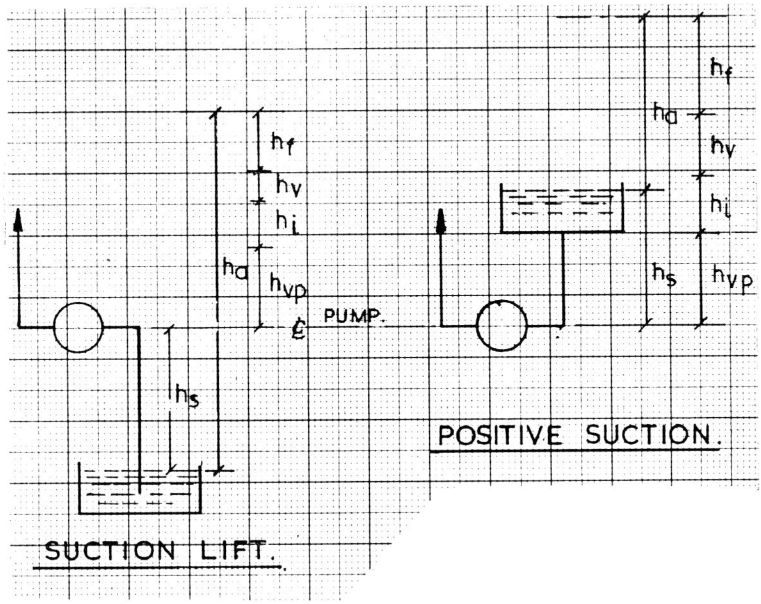

Fig. 2.8 Suction conditions

vaporisation does not occur at the impeller, the net sum of all these factors and atmospheric pressure must be greater than the vapour pressure of the liquid at the temperature concerned (see Fig. 2.8):

$$h_a \pm h_s - h_f - h_v - h_i = h_{vp}$$

where h_a = atmospheric pressure head
h_s = static suction head above or below pump centre-line
h_i = impeller eye head losses
h_{vp} = vapour pressure head of the liquid,

all measured in consistent units of length.

If one considers the algebraic difference between the left-hand and right-hand sides of the equation, the resulting value must be positive to avoid cavitation.

The equation may be re-written

$$h_a \pm h_s - h_f - h_v - h_{vp} = h_i$$

Closer scrutiny of the equation shows that all the terms on the left-hand side are items which the system designer has under his control, whereas h_i is an experimental value known only to the pump manufacturer, and varying with the quantity passed through the pump. Having allowed for all the left-hand side terms, the net result must be sufficient to overcome h_i or cavitation will occur. This net result is called the Net Positive Suction Head (NPSH), and when quoted by the system designer as a result of his analysis of the system, is referred to as the 'NPSH available'. Where quoted by the manufacturer as the known impeller losses with a safety margin, it is referred to as 'NPSH required'.

Both the NPSH 'available' and 'required' vary with the quantity being pumped, and it is customary for manufacturers to present the required NPSH as part of the pump performance curves (Fig. 2.9).

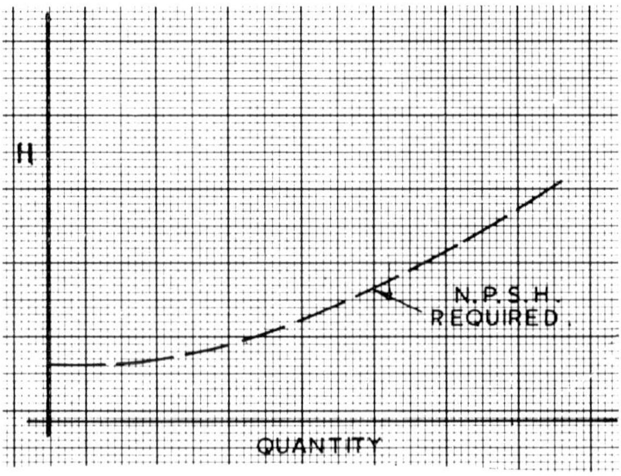

Fig. 2.9 NPSH required.

It is, however, of much greater value to the customer if the presentation is made in the form of a combined value of $h_a-(h_i+h_{vp})$, when quotations for pumps are called for. This merely re-arranges the equation into the form

$$h_a-(h_i+h_{vp}) = \mp h_s + h_f + h_v$$

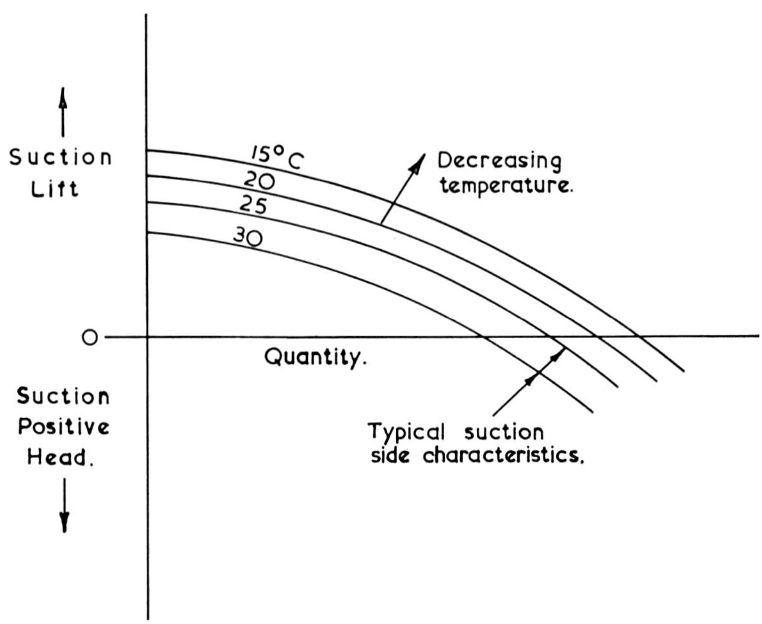

Fig. 2.10 Suction Conditions—Variation of temperature and suction head (See also Fig. 3.2).

taking suction lift as positive, which keeps all system physical losses together and enables the system to be studied against cavitation at various temperatures.

The left-hand side of the equation can now be termed the pump 'suction characteristic' and given in the form shown in Fig. 2.10, upon which the system designer can reproduce the characteristics of his system suction side as shown in Fig. 3.2. It will be noted that the warmer the liquid being pumped, the more sensitive the suction side is to cavitation occurring. This is due to the increasing value of h_{vp} as temperature increases. It should also be noted that the value of h_a drops by about 1 m for every 1000 m increase in elevation above sea-level, and that the effect of this decrease can be critical in elevated locations. 'Standard' designs need careful assessment therefore before they can be used at all elevations and in all kinds of climatic conditions.

SECTION THREE

PUMP AND SYSTEM COMBINED

(a) Duty points and combined characteristics

When a pump is matched to a system, that system's *Q-H* characteristic, superimposed on the *Q-H* characteristic of the pump, will cross at a point satisfying both system and pump characteristics. This is the stable point of quantity and head at which the pump freely operates in the system, and is called the 'duty point'.

It should be noted that the characteristics are the overall curves for both pump and system. In the case of the pump, any H value is the manometric total head developed across the pump. In the case of the system, all losses are included from entry to exit, and all differences of elevation taken account of algebraically. Typical combinations are shown in Fig. 3.1.

If the systems concerned are pumping cold water from a flooded suction through a short adequate suction pipe, then in all probability no further analysis need be made, and the pumps may be assumed to operate at the duty points shown.

If, however, there is a suction lift, a lengthy suction pipe, or elevated temperatures, it is dangerous to assume that all is as indicated, and the suction side conditions must be analysed to ensure that cavitation is not likely to occur.

To this end, the suction side static, friction and velocity losses are plotted and reproduced, as in the lower half of Fig. 3.2. Examination of the curves will now show that with system A the quantity q_A indicated by the overall duty point 0_1 will not be achieved if the temperature of the liquid is above 20 deg. C, since cavitation starts at that limit.

The most effective way of rectifying this condition is to change the nature of the suction lift. For example, if the head of water on the suction side is positive relative to pump centre-line, as indicated in system B, the suction side characteristic immediately makes available a quantity much greater than system A before cavitation occurs. At 20 deg. C this point is at q_B. The overall system resistance has however been lowered due to the decrease in overall head, giving the new duty point 0_2, tending to have a compensatory effect. This is most unlikely to have a more pronounced effect on quantity than radical changes in suction head, so that q_B is less than q and the system is satisfactory. If the suction condition is achieved by merely dropping the pump to give improved h_s, then there is no change in the overall head condition.

Similar changes can, of course, be achieved by increasing the size of the suction pipe to give a suction resistance as shown in the dotted curve A_1. It should always be remembered that any significant changes must be checked back on to the overall curves, particularly in a low head system if corrections are obtained by alteration of total head.

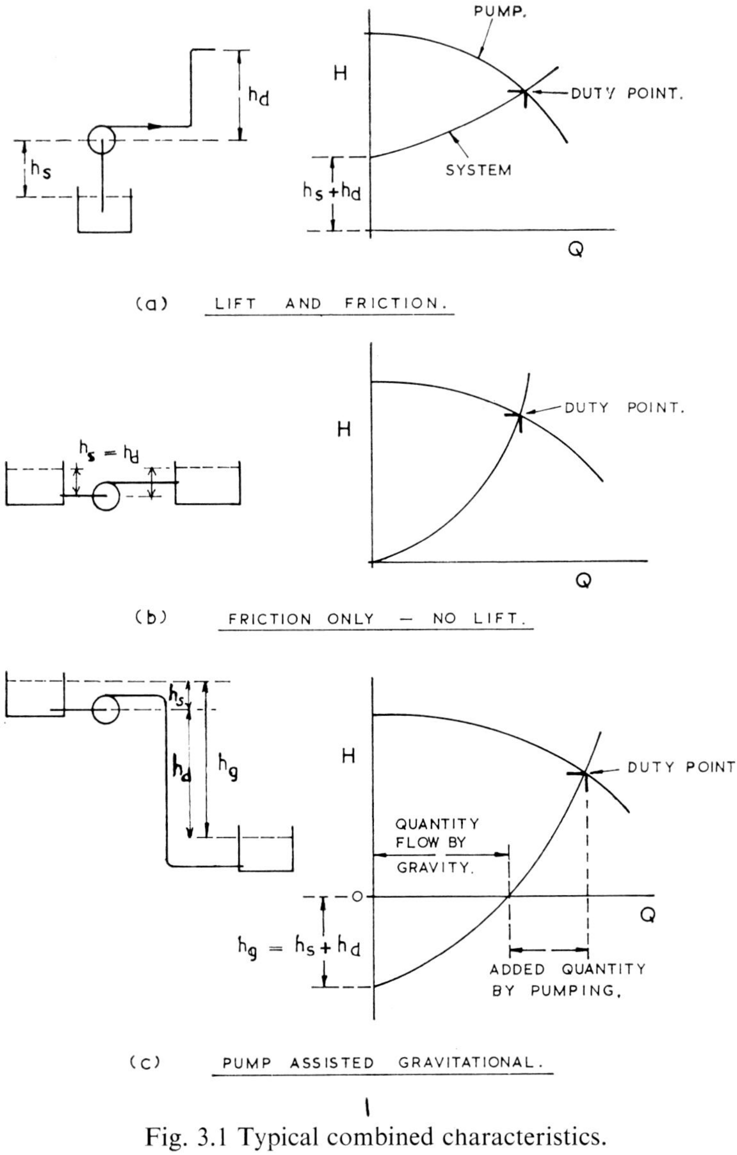

Fig. 3.1 Typical combined characteristics.

(b) Parallel and series operation

When pumps are operated in parallel, or in series, the head on the system and the quantity **are defined** by the point of intersection of the system resistance with the new **combined** pump characteristic. For series operation at any point value of Q, the **head** of that point is increased by the number of pumps in series. For parallel operation, for any point value of H, the **quantity** is increased by the number of pumps in parallel.

It is possible, as shown in Fig. 3.3, for an increase in capacity (or head)

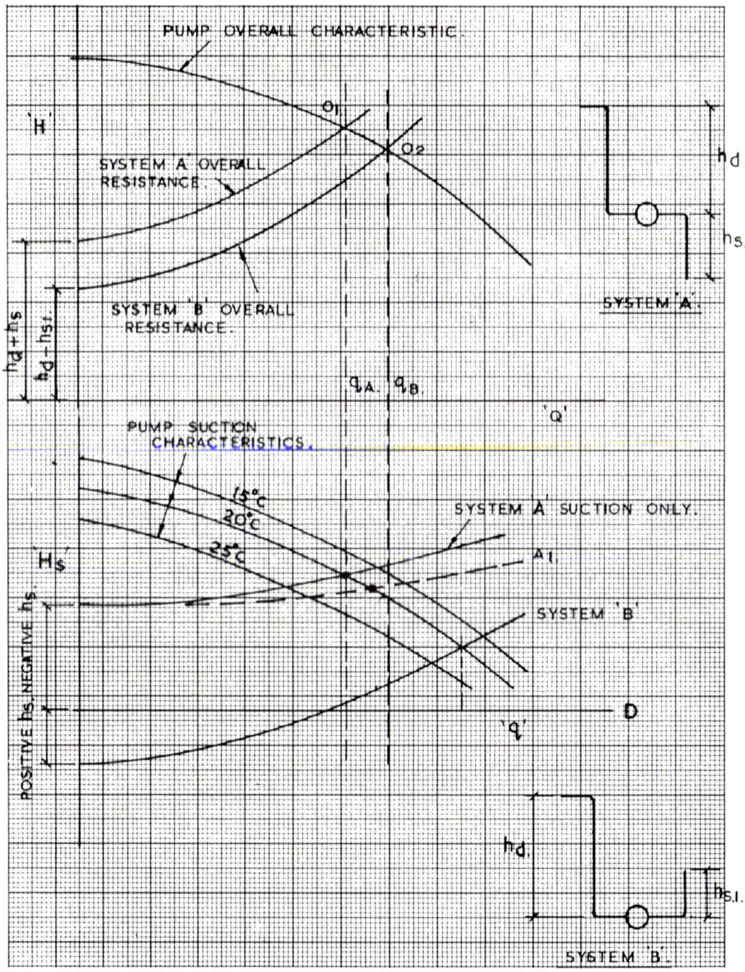

Fig. 3.2 Effect of changes in suction conditions.

to be obtained by either parallel or series working, and a decision on which to employ depends largely on the shape of the system characteristic. With a characteristic (a), a substantial increase in Q can only be obtained by parallel working, whilst for the (b) system resistance, the pumps have to be operated in series. The combined series and parallel curves cross at S, and if the new system characteristic lies to the left, series multiplication is indicated, and vice versa.

Generally, series operation should not be used, unless it is unavoidable, owing to questions of reliability in operation and difficulty of maintenance. If one pump needs maintaining, elaborate disconnections are necessary to keep the plant operating. In parallel operation, on the other hand, each pump can be valved off. It should be noted too, that in parallel the loss of

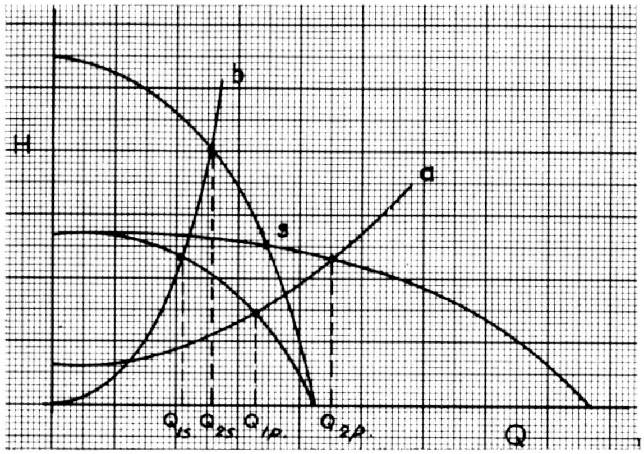

Fig. 3.3 Choice of series or parallel working.

one pump out of three does not mean a quantitative loss of a third. Conversely, the more pumps that are combined, the less efficient the multiplication becomes, in terms of pump usage, since the increase in quantity obtained relative to the *system* characteristic becomes less and less. From Fig. 3.4 it will be clear that:

$Q_2 < 2Q_1$ and $Q_3 \ll 3Q_1$.

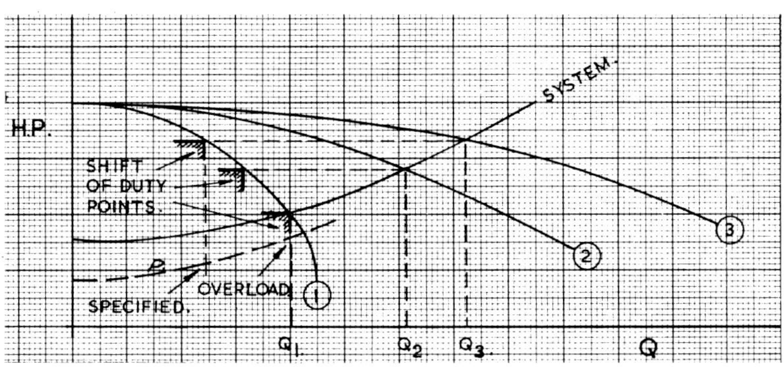

Fig. 3.4 Three similar pumps in parallel.

(c) Stability of pump curves

Stability is in essence the ability of a pump at any one head to produce only one flow condition. Not all centrifugal pumps are so endowed by their characteristic, notable exceptions being those of low specific speed and high impeller vane discharge angle. Such pumps tend to give a humped characteristic curve, of the type shown in Fig. 3.5, from which it will be noted that the 'closed valve' head is less than the maximum head of the characteristic.

All such pumps are potentially unstable, but it should be clearly understood that they may not actually show any sign of instability.

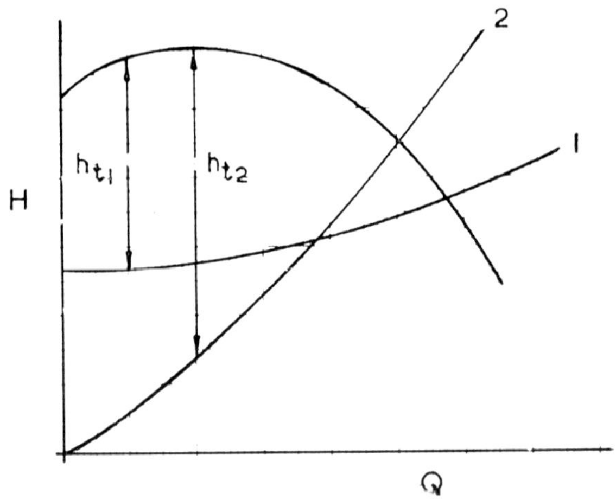

Fig. 3.5 Unstable pump characteristic.

Stability is a function of the pump with respect to the system in which it is to operate. The system and pump have to be examined together in two conditions: firstly, considering starting for both single pumps and pumps in parallel; secondly, considering the running condition of the system.

(i) *Starting conditions*

1. *Single pumps*

Referring again to Fig. 3.5, the system resistances against the pump characteristic are of two kinds: (1) which is a high static lift, low friction type, and (2), which is a high resistance type. On starting the pump, the duty point of each system is obtained by adjustment of the pump throttle valve, which is eventually fully open (except for any control margins) at the duty point.

Consider now System (2). The excess head to be dissipated in the throttle valve is the vertical distance h_{t2} between the pump curve and the system curve at any point of flow. It will be noted that in System (2), the degree of valve opening (i.e. the length of the throttling ordinate) decreases from 'no flow' to 'full flow', although the pump curve itself is humped. Conversely, with System (1), the distances h_{t1} after the initial valve opening reach a maximum at some point defined by the departure of the two characteristics, and then decrease again to nil at the duty point. Clearly, the action of the valve is then one of initial opening to a certain amount, of carefully closing and finally opening to the full. The pump is therefore considered to be unstable with respect to System (1), from which it should be noted that a humped pump characteristic used with a very flat system characteristic is always likely to result in an unstable starting condition.

2. *Two pumps in parallel*

Where two potentially unstable pumps are to be used in parallel, the conditions of stability become more complicated, and are outlined in Fig. 3.6. With one pump running, the duty point is at A, and to establish the final **system** operating point, with two pumps, the system resistance curve is produced beyond A, and the single characteristic doubled in terms of quantity for any head, until the two-pump characteristic is obtained. Where they cross is the two-pumps-in-parallel operating point B. The two-pump characteristic serves no further use than to establish point B, and may now be disregarded.

Returning to the sytem with one pump operating at A, the change in quantity and head (in terms of the system) when the

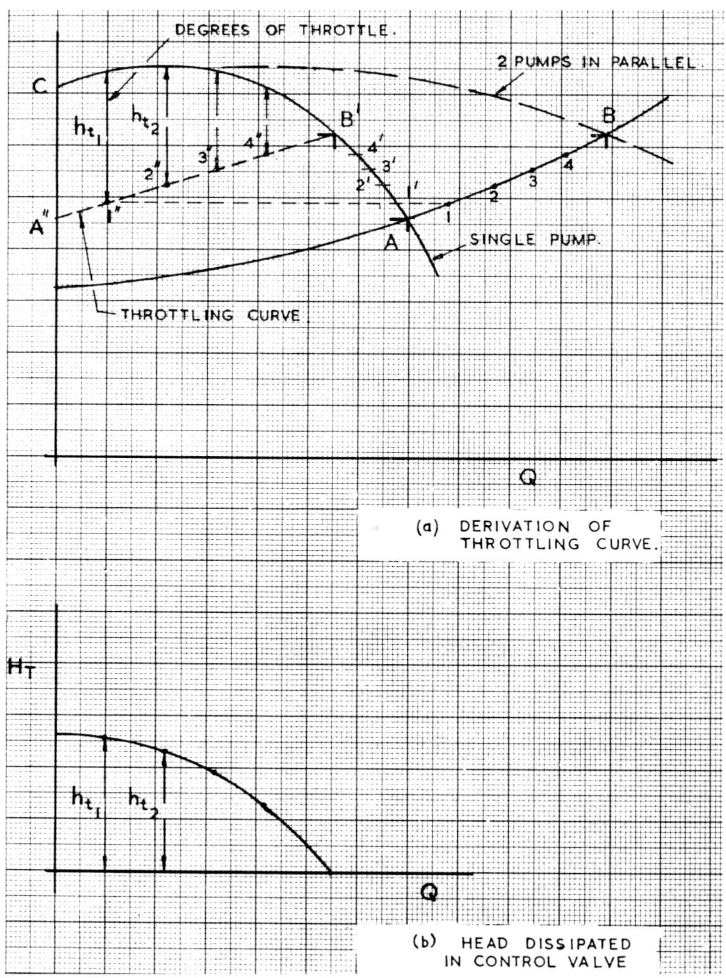

Fig. 3.6 Stability of pump starting.

second pump is brought on line is represented by a gradual movement from A to B passing through intermediate points 1, 2, 3 and 4. At the same time, both pumps are required to adjust themselves so that finally they are both operating at point B, each contributing one half of the system's quantity requirement of B at the correct head. To make this adjustment, the pump already on line must retreat along its characteristic from A to B′, at the same time as the pump being started advances along its characteristic from C to B′. In doing this, the first pump is progressively contributing less flow to the system as the second pump progressively contributes more.

Consider now any intermediate point in the process concerned. At 1 the first pump has retreated to point 1′ on its characteristic in order to gain the necessary increase of head. A quantity 1 – 1′ must therefore be obtained from the second pump at this point. Further, it must be delivering this quantity against the head requirement of point 1. Point 1″ is therefore established a distance from the origin equal to 1 – 1′ quantitatively and at the head corresponding to the system resistance of point 1. Similar points are established corresponding to 2, 3, 4 and the interconnecting curve drawn. It will now be appreciated that, in bringing the pump on-line, the excess head to be throttled off in the second pump's control valve is the vertical ordinate above each point on the curve consecutively – at point 1 the throttle is ht_1, etc.

To examine the pump for stability at starting, the requirement is now that the ordinates so found should become progressively smaller as the pump approaches B′. When plotted on a separate graph of Q against throttle head, if the curve is progressively diminishing, a stable condition obtains. If this curve is humped, then the pumps are unstable in parallel in the system concerned.

This does not necessarily mean that they cannot be started in parallel, but that such a start is a carefully controlled affair with the characteristic of the control valve as outlined in Fig. 3.6 (b).

In the example given, a potentially unstable pump is shown to be quite stable when brought on line in that particular system against a similar pump already working. To bring the first pump on line originally may, of course, prove to be an 'unstable' case, and should be checked independently as previously outlined. Information on the selection of control valves can be obtained from E.E.U.A. Handbook No. 32.

(ii) *Operating conditions*

Despite the fact that a potentially unstable pump may be brought on line stably, its operation condition may prove in itself to be unstable. To investigate this aspect, consider Fig. 3.7, which shows a system having a variable static lift and two free water surfaces. At one condition of lift, the system curve 1 is appropriate and the pump may well be stable. If, however, the demand and head on the system changes, so that system curve 2 applies, it will be apparent that the movement has carried the pump duty point beyond the 'hump' in the characteristic. As soon as the head in the system has increased to the maximum pump head D, then the pump can no longer provide the necessary pressure contribution, and a tendency to instability occurs. The pump tends to return to the no flow condition and, as a result, the resistance of the system drops, consequently causing the pump to increase head and quantity delivered again in trying to arrive at a balance. An oscillation is therefore set up, and the condition of operation becomes unstable.

Such conditions are most likely to occur when the system has free water surface, which may (or may not) be under pressure. Pressure surfaces may even accentuate the oscillation. It is therefore of considerable importance in systems such as a boiler feed installation to be operating in a stable state, with pumps that have fundamentally stable characteristics.

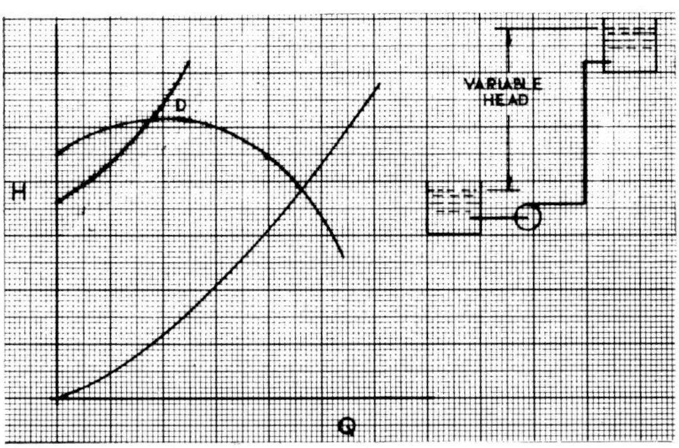

Fig. 3.7 Unstable pumps: shifting system characteristic.

(d) Maximum power conditions

Where a system is designed specifically for operation in parallel with either one, two or three pumps in operation, careful analysis is required to ensure that the pumps are not overloading at the single pump duty point. Fig. 3.4 shows three such pumps and the transferred duty points at which single, double or triple working occurs. If they are specified for the maximum head condition only, the lower head point may cause overloading of the motor, as shown. In such cases, all three duty points should be specified, and the power characteristic should be of the non-overloading type.

(e) 'Effective' pumps and branched systems

The characteristics so far drawn have been those of the complete system measured against the pump *at the pump*. There may, however, be a branch from the system at point A, Fig. 3.8 (a) and the effect of the pump on the branch line may be required, when pumping a combined quantity into both legs of the system. It is convenient, therefore, in analysing quantities passing through both system divisions, to imagine the pump as it would be if it were located at A. Consider now the characteristic of the pump plotted against the system's characteristic from D to A, taking h as the elevational difference and h_f the friction losses at any quantity q. The pump's effect in any of the branches is as if the h_f losses had been deducted from the pump's performance, and so the pump's characteristic measured at A will be the reduced dotted line in Fig. 3.8 (b). The branch conditions may now be considered on the simple basis that what enters one branch is not available to the other. In Fig. 3.9, 1 is the system resistance of AB and 2 that of AC. When the duty point of the effective pump is at X, the quantity passing through system 1 will be q_1, and through system 2 will

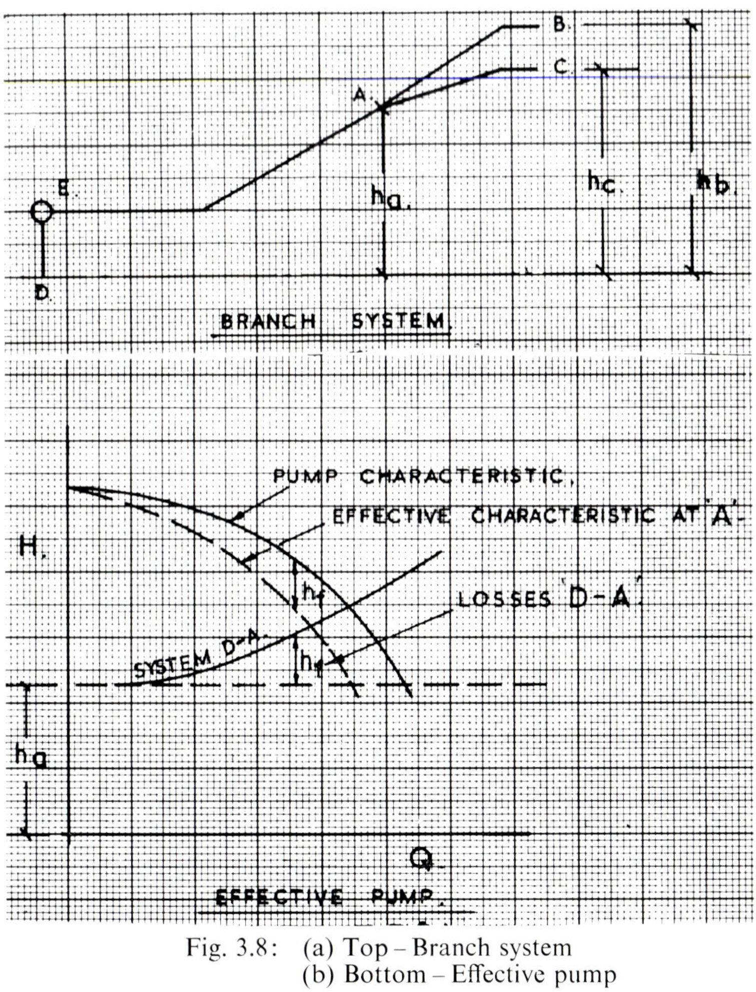

Fig. 3.8: (a) Top – Branch system
(b) Bottom – Effective pump

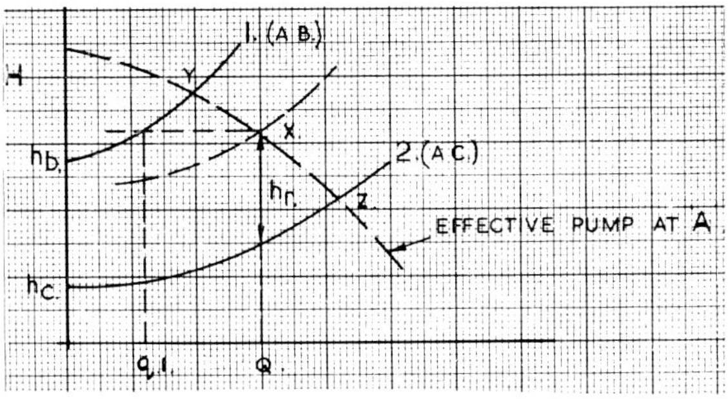

Fig. 3.9 Branch flows.

be $(Q-q_1)$. Clearly, at this point, system 2 will be throttled at C to give a residual pressure h_r.

Also, by examination, if point X moves to Y, then Q and q_1 coincide, and there is no flow in system 2 (AC).

Equally, if system 2 is unthrottled, the head (at X) will fall below B and no further flow will occur in AB. Final de-throttling will result in eventual balance on system 2 alone at Z.

Take the case when point Z is above h_b, as indicated in Fig. 3.10.

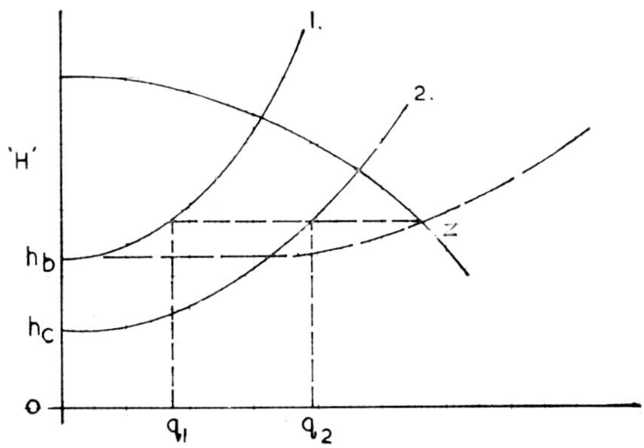

Fig. 3.10 Branch flows

The two characteristics must now be combined to give the dotted line. The condition for no flow in 2 is the same as before, but 2 can no longer starve 1 entirely of flow. The combined open valve condition is at Z, giving q_1 and q_2 respectively as the flow in each pipe.

Another example of the use of effective pump characteristic is the insertion of a pressure-reducing valve in the system (Fig. 3.11). Let us assume that system A B C D is protected by a pressure-reducing valve at E. The effective pump at B is required in order to analyse branches BC and BD. Clearly, no pressures of the pump exceeding the pressure-reducing valve setting will be felt beyond E, and the analysis, as before the pump, is superimposed with the cut-in pressureof the valve plus the losses from the

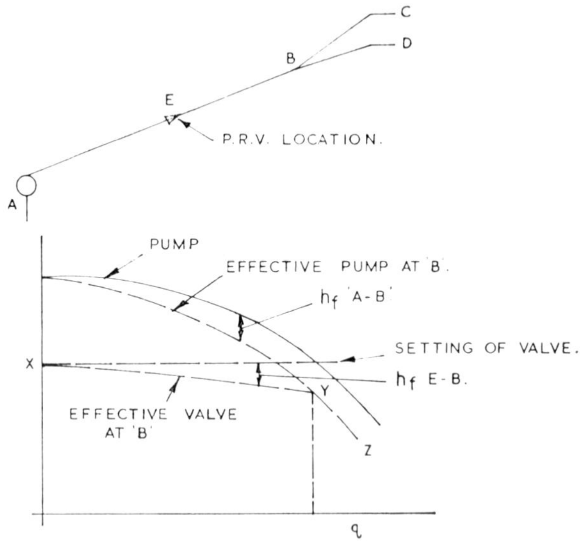

Fig. 3.11 Effective characteristic with pressure reducing valve.

valve to point B, since the effect of the valve at B is reflected through the pipework EB. Fig. 3.11 shows the net effect and it is seen that the pressure-control valve is the effective control for all quantities up to q when the pump controls the system. Flows in branches B and C must therefore be judged against the effective net curve XYZ.

(f) Modifications of duty

Variations of Q and H with speed N have already been discussed in Section Two. The need to use these relationships in practice is frequent, since any installed system tends to become modified, as changes in the use of the plant become necessary.

Typical changes of this sort are shown in Fig. 3.12, where the original duty point is A. An increased product demand, coupled with additional pipework, means a new duty point at B, and a reduced demand on the same pipe system is indicated at C, where a worthwhile power saving may be experienced. In both cases, change in speed or in impeller diameter are indicated.

Changes in speed may well mean changing a motor, and, coupled with the fact that this is usually an expensive modification, precise changes in duty or speed cannot always be met in this way. It is far more likely that such changes to new duty points are solved by changes in impeller diameter: a decrease to meet case C and an increase to meet case B, in Fig. 3.12.

These changes can be calculated as previously indicated for radial conditions on the basis that

$$\frac{d}{d_1} = \frac{Q}{Q_1} = \sqrt{\frac{H}{H_1}} = \sqrt[3]{\frac{P}{P_1}}$$

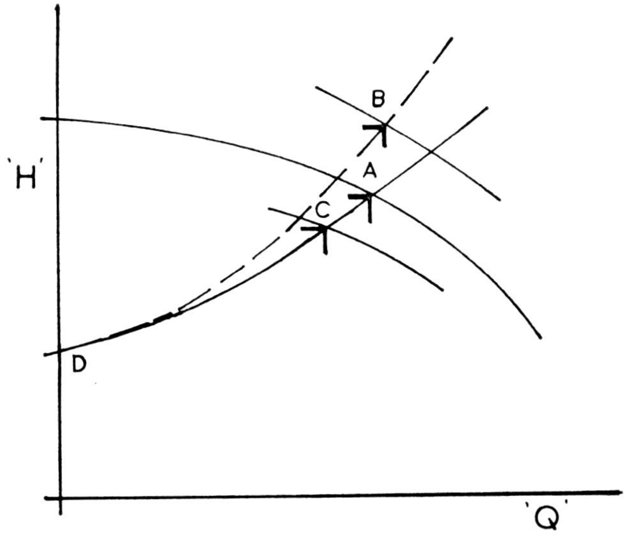

Fig. 3.12 Modifications of duty.

but as the impeller is reduced, these rules become approximate due to changes in efficiency and changes in theoretical design assumptions.

The graphical analysis shown in Fig. 3.13 largely eliminates trial and error methods in assessing a revised impeller. The new duty point X is shown in relation to the head-capacity curve of an existing impeller. Any capacity Q_Y is then chosen beyond Q_X and the corresponding head to H_Y is calculated from

$$H_Y = H_X \left[\frac{Q_Y}{Q_X}\right]^2$$

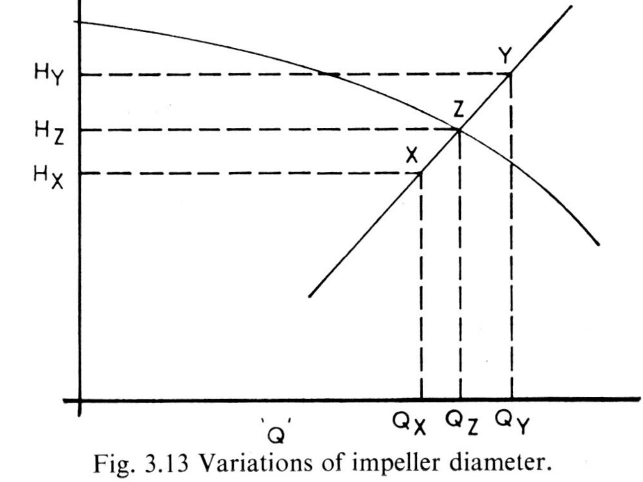

Fig. 3.13 Variations of impeller diameter.

Point Y should lie beyond the known Q-H curve, so that points X and Y can be joined, intersecting the known Q-H curve at Z. The required impeller diameter ratio is then

$$\frac{Q_X}{Q_Z}$$

Similarly, an increase may be calculated. The line XY lies close to the equal-efficiency line through X, and Y and Z have approximately the same specific speed.

The above applies to radial pumps only, and although in principle the same technique can be applied to mixed flow pumps, for constructional reasons not so much cutting of the impeller can be done.

With axial flow pumps, the capacity will vary with the swept area of the impeller, and the head at any point on the impeller varies with the square of the diameter at that point. Reducing the diameter of an axial impeller would require a new ring or new casing, a solution not often used in practice. Variations are obtainable, however, by altering the angle of attack of the propeller blade.

For radial or mixed flow pumps, the question of increasing the impeller diameter at some future date should always be borne in mind when initially buying the pump. One should carefully consider whether any pump should be bought whose frame size will not take a reasonable increase in capacity at a later date. It is always advisable to have 5% at least in hand, to bear increases within any casing, and usually $\pm 15\%$ are acceptable variations, limited on the one hand by the casing size and on the other by loss of efficiency.

Where there is a known possibility of an increase in diameter of impeller, it is also very important for the motor to be capable of dealing with the increased pumping quantity when the impeller is changed. The capacity of the driver must in any case be checked when an impeller is increased in diameter, to ensure that it is not overloaded.

(g) Pumping viscous liquids

When a pump has to move a viscous liquid, the characteristics will be different from those for water. Fig. 3.14 shows the characteristics for water in continuous lines and for a viscous liquid in dotted lines.

The greater the viscosity, the higher the power absorbed by the pump, and this may assume such proportions that the use of a centrifugal pump becomes uneconomical. If this is the case, a positive displacement pump (e.g. screw pump) may have to be used.

There is no definite limit to the viscosities of liquids pumped above which centrifugal pumps should not be used. The choice between centrifugal and positive displacement pumps depends on a number of factors, such as capacity and head, initial cost of equipment, running costs, maintenance costs, running time, etc.

Positive displacement pumps are beyond the scope of this handbook,

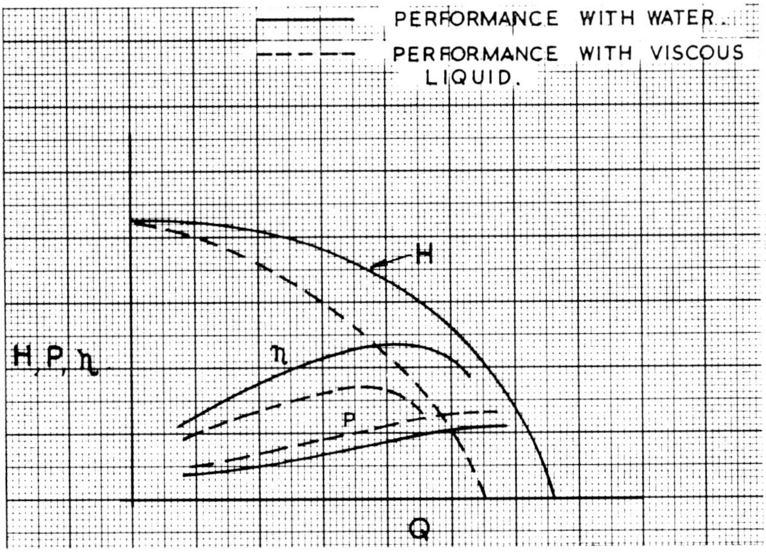

Fig. 3.14 Viscous liquids.

but figures on power consumption may readily be obtained from pump manufacturers' catalogues.

With the aid of Fig. 3.15, an estimate can be made of the power absorbed by a centrifugal pump when pumping a viscous liquid. It should be noted that the results are only a rough approximation. Especially with pumps which, on water, already have a low efficiency (low capacity pumps), the calculations may yield too high a value for power absorbed.

The graph can be used in two ways:

- For calculating (starting with known characteristics on water) the head/capacity point and power absorbed at maximum efficiency when pumping a viscous liquid.
- For calculating the size of pump required for pumping a viscous liquid.

In the example below, using Fig. 3.15, a pump has been considered which, at maximum efficiency point, has a capacity of 250 m^3/h (1100 U.S. gal./min.) against a head of 43 m (140 ft) when pumping water. The maximum efficiency is assumed to be 85%.

On water, the power absorbed at maximum efficiency point is:

$$P = 9{\cdot}807 \times \frac{250 \times 10^3 \times 43}{60 \times 60 \times 0{\cdot}85} = 34 \text{ kW}$$

To calculate the performance of this pump and the power absorbed when pumping a viscous liquid, the following liquid is considered:

specific gravity = 0·9
viscosity = 440 cS (2000 SSU)

To arrive at the new duty point, correction factors have to be applied to

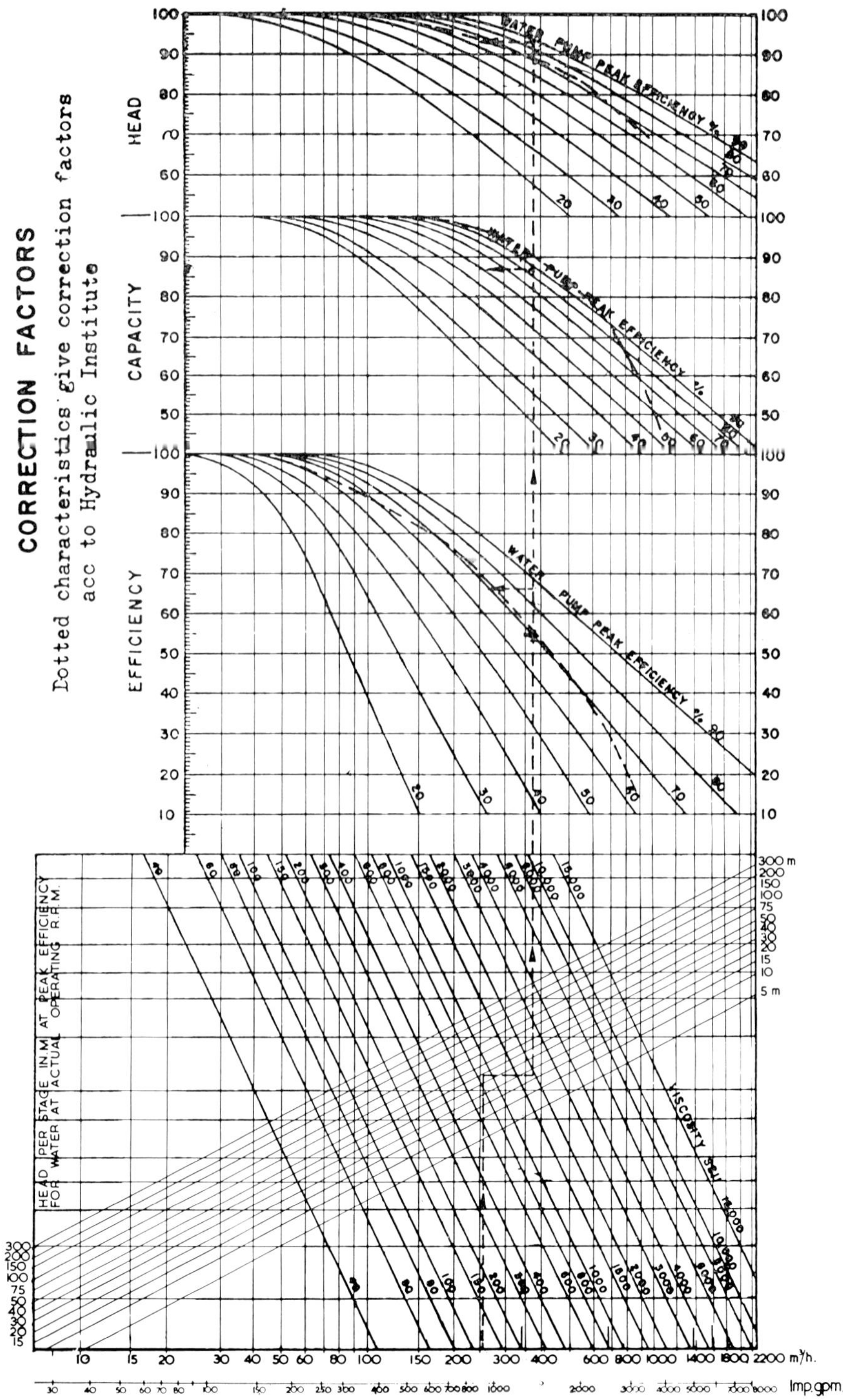

Fig. 3.15 Pump viscosity correction chart.

head and capacity. In addition, a correction factor has to be found for the efficiency. It is evident that the efficiency will be decreased, since capacity and head will be lower and power lost because of the internal friction of the liquid.

Using Fig. 3.15, the example can be developed as follows:

- Project capacity point, 250 m^3/h from the base of the graph vertically upwards to intersect the 43 m line.
- Project horizontally to the 440 cS line.
- Project vertically upwards to intersect the correction factor curves for 80% efficiency.

The correction factors thus found are:

for head: 0·94
for capacity: 0·87
for efficiency: 0·66

The new duty point is then:

Head $= 0{\cdot}94 \times 43 = 40{\cdot}5$ m
Capacity $= 0{\cdot}87 \times 250 = 217$ m^3/h

The efficiency at this point (assumed to be the maximum efficiency) is $0{\cdot}66 \times 85 = 56\%$

On this viscous liquid, the power absorbed at maximum efficiency point is:—

$$P = \frac{9{\cdot}807 \times 217 \times 0{\cdot}9 \times 10^3 \times 40{\cdot}5}{60 \times 60 \times 0{\cdot}56} = 38{\cdot}5 \text{ kW}$$

It will be noted that there is only a slight increase in power for pumping the two liquids. But, in the case of the viscous liquid, the capacity and head are lower and so is the specific gravity. A screw pump might have an overall efficiency of 65% on this viscous liquid and the power absorbed would be only 21 kW.

The example given in Fig. 3.15 can also be used to estimate the size of a pump as regards capacity and head on water when the duty on viscous liquid is known. The graph may be used for this purpose, since the results are an approximation in any case.

Assume a viscous liquid with the same characteristics as above, but duty point 250 m^3/h and 43 m. The pump should be able to pump water at a capacity of $\frac{250}{0{\cdot}87} = 287$ m^3/h against a head of $\frac{43}{0{\cdot}94} = 46$ m., assuming a pump efficiency on water of 85%. A pump of these characteristics will be able to perform the required duty on the viscous liquid.

The power absorbed at duty point when pumping the viscous liquid is:

$$P = \frac{9{\cdot}807 \times 250 \times 0{\cdot}9 \times 10^3 \times 43}{60 \times 60 \times 0{\cdot}56} = 47 \text{ kW}$$

Here again, a screw pump with 65% efficiency will require only 41 kW Fig. 3.15 is one of a number of graphs designed to calculate pump characteristics when pumping viscous liquids. The Hydraulic Institute (USA) use an identical graph, as far as the lower part is concerned, but, for the correction factors, the efficiencies on water have not been taken into account. (The dotted lines are used for correction factors). The present graph includes the peak efficiencies but, as mentioned above, they are not always applicable at the lower efficiency values.

It should be noted that if a two-stage or a multi-stage pump is considered, the head **per stage** should always be used in the calculations.

(h) Induced surges in pipelines

The analysis of surge conditions in a pipeline is complicated and time-consuming, and falls generally outside the scope of this handbook. However, the reader should be aware that his choice of pump may affect surge conditions in the associated system, and from this point of view the salient factors will be considered. For detailed information, the reader is referred to Parmakian, Stepanoff and others.

The surges concerned here are often referred to as water hammer, and should not be confused with the term 'surge' as applied to variation in flow, such as occurs in the conditions outlined in Section Three (c) when pumps with unstable characteristics are operating in unfavourable circumstances.

Surge is caused when the steady state of flow in a pipeline is disturbed for any reason. It is a transient condition lasting until a further steady velocity state (or state of rest) has been established. The degree of disturbance depends on the rate at which the velocity is increased or decreased, instantaneous changes causing high surge, while slow changes cause little disturbance of prevailing conditions.

If a sudden change occurs, such as a valve closing, the fluid behind the gate tends to build up in the restraining pipeline. The fluid is largely incompressible, and a shock wave is initiated which travels along the pipeline from the point at which the valve is located. This shock wave is reflected from the free surface of the fluid and travels back again to the valve. During the period of valve movement, normal pressures in the pipeline are modified by the algebraic sum of the transient shock waves emanating from the valve and those returning to it. Their severity depends on how quickly the valve is closing, the length, material and thickness of the pipe, the compressibility of the fluid and rigidity of the pipe mountings. Upstream of the valve, the pressures are generally increased relative to the previous steady state pressures, and vice versa downstream.

In the same way that a valve closure changes velocity of flow to give rise to such surges, so also does any obstruction suddenly introduced into the pipeline. This is precisely the case when a pump, normally operating with manually-controlled discharge valves, loses its power supply. The pump immediately slows down and acts as a brake on the system. The effectiveness of the 'brake' depends on the inertia of the pump, motor and other rotating parts. The resulting surge wave will depend on these inertia forces and the rate of change of the pump's characteristics as it slows down.

The tendency is always for a manufacturer to build to the lowest cost, and this must mean that the weight of moving parts is normally cut to a minimum. The run-down time of a pump on failure tends, therefore, to become small, and especially if a long pipeline with high velocities forms the system, then that system should be thoroughly checked.

On loss of power, the flow continues away from the pump and there is an immediate downsurge at the pump discharge which may, due to valve closure characteristics, become a pronounced upsurge later. Conversely, if the pump is subject to a long suction pipeline, the characteristic of surge on the suction side will be an immediate upsurge at the suction flange, with possibly a downsurge later as the valve closes. Eventually, the pressures on both sides return to the static pressures obtaining, a typical trace being shown in Fig. 3.17.

It should be noted, then, that the pump and valve should always be considered together, and that a suitable combination is one which reduces the maximum upsurge or downsurge to a minimum acceptable value.

In Fig. 3.16, if a pump fails, the fluid will eventually drain back through

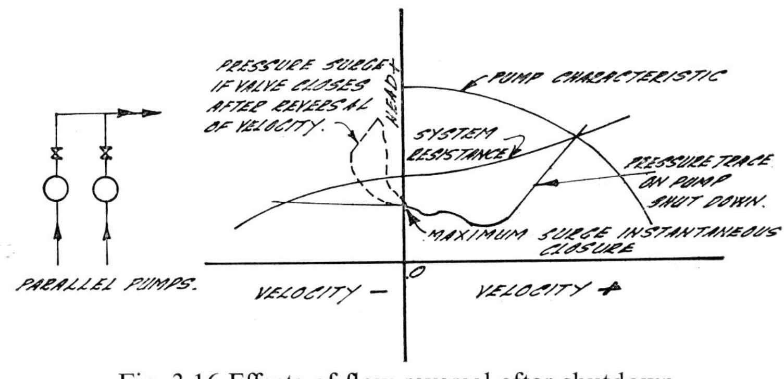

Fig. 3.16 Effects of flow-reversal after shutdown.

the pump, and in these circumstances the best time to close a valve is when this reversal is about to take place. Fast closing of a valve in such circumstances is not only acceptable but desirable, since leaving closure too late will meet increasing velocities in reverse. Surge in a system when the valve is closed at no flow would be confined solely to the effects of the pump running down, and this is a function of the inertia of the pump and its working head. The longer the run down time, and the greater the inertia, the less will be the resulting surge. It is, therefore, sometimes a cure to a surge problem to introduce a suitably-sized flywheel between the pump and motor.

Air vessels are often introduced into a system on the discharge side of a pump to minimise surge, advantage being taken of the compressibility of air to absorb the immediate shock effects by change of volume.

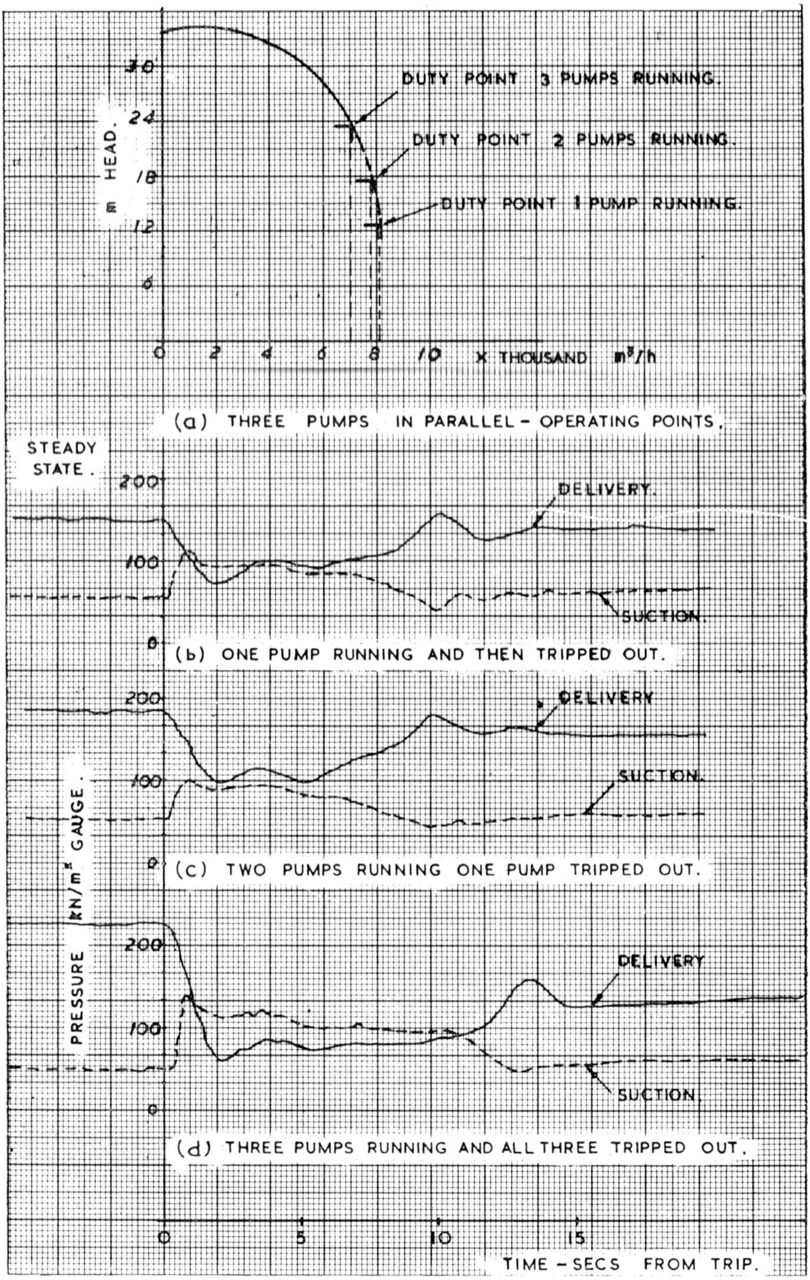

Fig. 3.17 Surges on power failure of parallel running pumps.

An example where precautions at the pump discharge have little effect is that of a 'loading line', where bins or cans are filled with instantaneous closure valves controlling the filling operation. In such cases, a by-pass line near the filling process can reduce shock and noise considerably (Fig. 3.18).

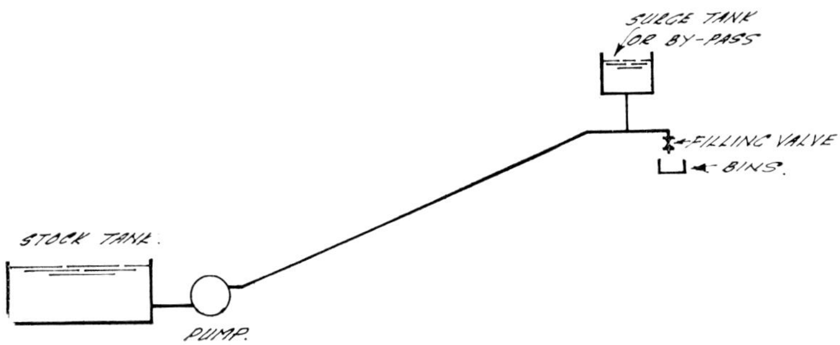

Fig. 3.18 Surge control at loading lines with quick-closing valves.

Perhaps the worst effect of surge occurs when the downsurge caused reduces pressure in any part of the system to below the vapour pressure, and the water column separates as a result. Considerable damage can occur when the system recovers and the two water columns are rejoined. This can happen when the syphon principle is used in part of a pumping system (Fig. 3.19). The hydraulic gradient is below the top of the pipeline

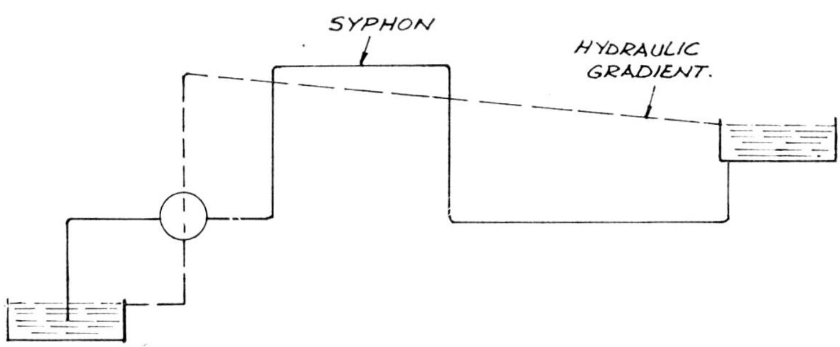

Fig. 3.19 Syphon on pumping line.

in normal working conditions (for example, where a pipeline is taken over other machinery or a road crossing perhaps) and when a downsurge occurs on pump failure the further loss of pressure may be sufficient to cause water column separation.

A similar effect is seen in Fig. 3.20 in the normal operation of a system. The pump is operating in conjunction with a non-return valve located at a higher point. If the pump is stopped, the water in the system is held in check by the non-return valve, but behind the valve the water level has dropped, leaving a partial vacuum. This acts precisely as water column separation when the pump is restarted, possibly fracturing the bend.

In general, then, any pump of low inertia, operating in a system with a long suction or delivery line, and with high velocities, is likely to cause a surge problem which it would be wise to examine.

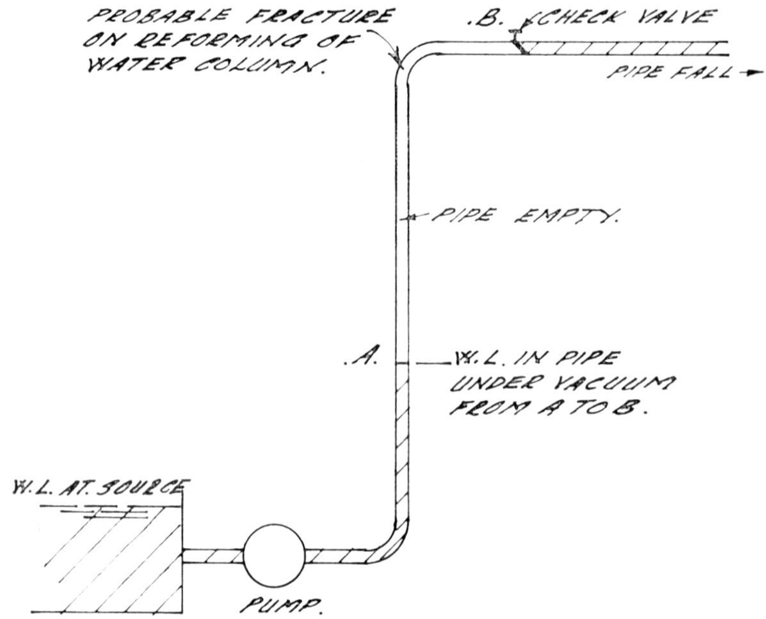

Fig. 3.20 Dangerous positioning of check valve.

(j) Effect of waterborne solids

Waterborne solids affect pump system design parameters in two ways: the specific gravity of the slurry has to be taken into account and the velocity of flow has to be sufficiently high to keep the suspended particles in motion.

The velocity required will generally depend on the specific gravity of the granules of the slurry, and should be not less than five times the free-falling velocity of the granule in the conveying fluid. Much higher velocities may be required, however, to prevent coagulation or settling, depending on the concentration of the slurry. A minimum velocity of about 1·6 m/s will normally prevent sedimentation, but much depends on the nature of the slurry concerned.

In terms of pumping power, the density to be considered is that of the slurry, not that of the parent liquid. The same applies when calculating the Reynolds Number (R_e), but the viscosity may be taken as that of the conveying fluid if the solids are in suspension.

SECTION FOUR

ECONOMIC CONSIDERATIONS

(a) Components of total pumping cost

The final aim of any pump user should be to attain the lowest possible total cost for pumping. This can only be achieved by considering all relevant aspects of the whole system – pumps, piping and ancillary equipment – before any part of the system is selected.

The total annual pumping cost is made up of the following components:

- Procurement cost for entire system, expressed as an amount paid annually during the useful life of the plant. This covers the cost of purchasing, installing and commissioning the whole plant, with interest charge on capital employed.
  It is here called the 'annual capital cost'.
- Energy cost for driving the pumps.
- Labour charge for plant operators.
- Maintenance cost for entire system.

Items (ii) and (iii) together, being roughly proportional to the annual running time, may be called the 'running cost'.

(b) Matters affecting design

In considering a complete new pumping system, the designer is faced immediately with choosing a pipe diameter somewhere between the extremes of small diameter (high velocity, high friction loss, high pump power) and large diameter (low velocity, low friction loss, low pump power).

It is generally necessary to make calculations based on several pipe diameters, to establish the one to give minimum total annual pumping cost. In doing that, the following circumstances should be considered.

(i) *Rate of flow and static head*

Throughout this section it is supposed that these are established quantities given to the pumping system designer, and that no overall economies in a project containing an ancillary pumping system can be gained by altering the static head or flow rate of the pumping system.

This is a matter for consideration by the engineer having overall responsibility for the project, who should, so far as is possible, ensure that the whole plant – not just the pumping system part of it – is designed with due attention to overall economy.

(ii) *The useful life of the plant*

This is the number of years over which it is expected that the pumping plant will operate at approximately the conditions for which it is being designed, without modifications of greater extent than are normally regarded as maintenance requirements. It is usually justifiable to assume that the plant value will be negligible at the expiry of this period.

(iii) *Annual operating time*

The estimated equivalent number of hours per year for which the plant will be in full-load service. This quantity, multiplied by the full-load hourly energy requirement, gives the annual energy requirement. Proper consideration of operating time is important. In a system with low operating time this leads to smaller pipe sizes, smaller capital costs, and higher pump power, in comparison with a system of the same pumping capacity designed for high operating time.

With continuous working the plant may operate for nearly all of the 8760 hours in the year. With a typical single shift 5-day week the annual time would not be more than about 2000 hours. With intermittent operation, the annual time could be lower than 500 hours.

(iv) *The nature of the fluid to be pumped*

If the corrosive nature of the fluid warrants the use of expensive materials, an economically-designed system will usually have a smaller pipe diameter and higher pump power than would otherwise be chosen.

With a fluid containing much abrasive material, considerations of maintenance will impose a top limit on velocity.

(v) *Security of the system*

The capital cost is greatly affected by the inclusion of features such as doubly-secure power supplies, heavy instrumentation and fully automatic control. Careful thought must be given if the need to ensure high reliability, availability and efficiency of the pumping service appears at first to justify such features. If it does, they must be included in calculations involving capital cost.

In many cases, it is necessary to provide standby pump capacity to enable the system to work at full design duty when one pump is being overhauled. There is then a choice of having one running pump and one standby, or a number of smaller running pumps and one standby. Consideration of the physical size, power and cost of the units (and of the part-load performance) will usually show the total number to choose.

In cases where a breakdown would cause little consequential cost or inconvenience, provision of a standby pump may not be justifiable; this implies that replacement parts would be available with acceptable delay.

(vi) *Control of pumps to give efficient operation*

Annual running costs can in some cases be considerably decreased by careful attention to the design of the control system, to avoid long periods of operation at low pump efficiency.

This can be specially important where a pump feeds into a storage tank from which abstraction is intermittent.

In many such cases, total annual cost can be reduced by choosing a tank of sufficient capacity for the pump to be stopped and re-started by automatic control, rather than kept running under the control of a float valve at intermediate flows.

(vii) *Possibility of increase in demand*

In some cases, it is clear from the outset that there will be a planned increase in demand during the useful life of the plant, whereupon it may be economic to install at the start such larger piping as is necessary to accommodate the ultimate demand, though all the pumps may not be installed at that time. In other cases, a later demand increase may be foreseen only as a possibility. The degree of allowance to be made for this in initial design would depend on the magnitude of the possible increase, and the practical difficulties associated with later modifications if no allowance is made in the initial installation. Even in cases where there seems to be no likelihood of an increase, the possibility should be borne in mind.

Calculations may show, for instance, that the total annual pumping cost is nearly the same for two pipe sizes, as, for example, Case D plotted in Fig. 4.1, where one might choose 356 mm diameter in preference to 305 mm because it would permit a later demand increase at lower additional cost for new pumps and extra power. The initial capital cost would be about 20% greater than for 305 mm, but, provided the system remains in use for not less than the estimated time, this is of practically no consequence.

(viii) *Cost of installing power supply*

In most cases, electric motors are the best choice for driving pumps, but much depends on the pump siting in relation to the nearest existing sub-station. The cost of providing electrical supplies for a pump-house in open country may be very high, and in such a case diesel engine drive may be the best choice.

(ix) *Cost of energy*

Ideally, the unit energy cost value employed in design calculations should be the average value over the life of the plant: that is, present cost with an adjustment for future changes. For electricity purchased from public utility organisations, future trends can only be estimated by extrapolation. Fig. 4.2 shows average British prices for supplies to industrial premises from 1950 to 1967.

Where the electricity is generated by the pump user's pass-out or

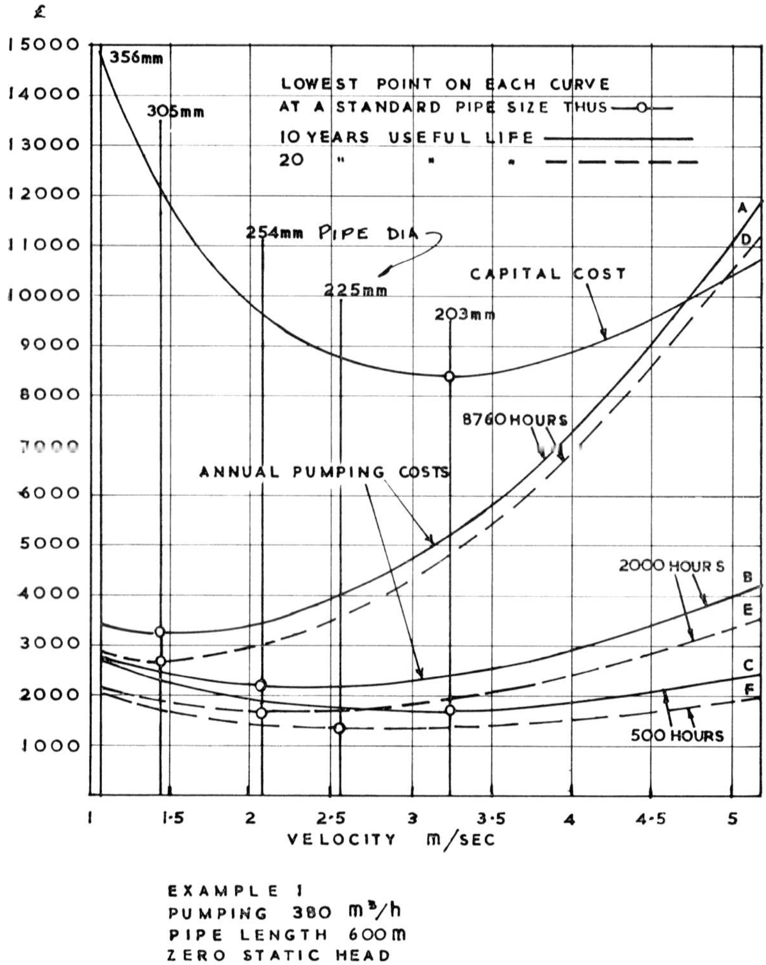

(Example 1 refers to water)

Fig. 4.1 Cost variation: no static head.

back-pressure steam turbine plant, calculation of energy costs is more complicated. Here it is necessary to use marginal (not average) generation costs in the pumping system design calculations. In evaluating marginal generation costs, it is sufficiently accurate to take account only of the additional boiler fuel consumed, since usually the generation of some extra power does not incur additional labour, replacements or overhead costs.

(x) *Cost of civil engineering work*

The cost of civil engineering work may be considerably affected by the choice of pump type. For example, vertical mixed-flow pumps might

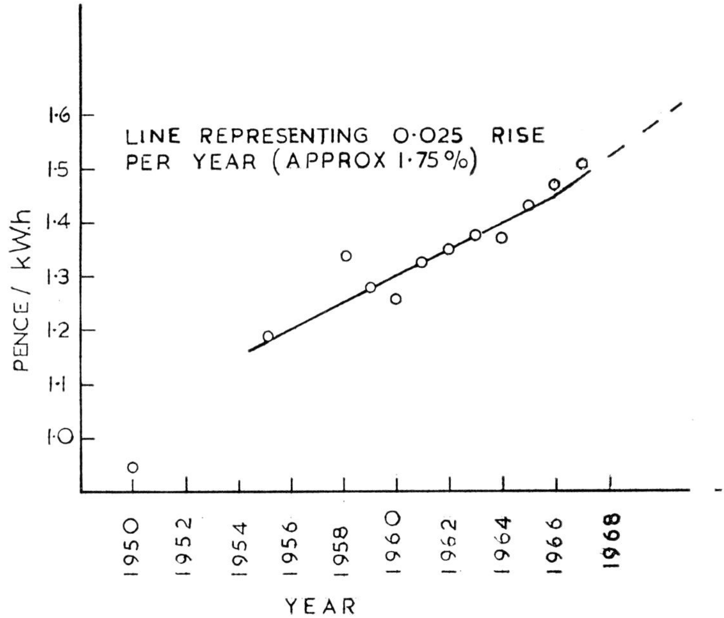

Fig. 4.2 Average price of electrical power.

Average price of energy sold by Electricity Boards in Britain to industrial premises. From Ministry of Power Statistical Digest 1967.

be considered as an alternative to horizontal centrifugal pumps for a particular application, but there would probably be a considerable additional civil cost due to the dry well of the horizontal pump installation. Where the designer has to choose between overhead and underground piping, civil engineering costs are among the important decisive factors.

(xi) *Relative importance of capital expenditure and total pumping cost*

If available capital is restricted, one might be justified in choosing a pipe size smaller than that giving minimum total annual cost, in order to reduce capital expenditure. Example 1 in Section Four(d) (costs plotted in Fig. 4.1) illustrates this. Minimum capital expenditure is obtained at 203 mm dia. With the plant operating continuously, minimum total annual cost is obtained at 305 mm dia. By choosing 254 mm instead of 305 mm diameter, the capital expenditure would be reduced from about £12,000 to £9,500, but the total annual cost would be increased from £3,200 to £3,400 on the basis of 10 years' life.

With decrease in annual operating time, the pipe size giving minimum total annual cost approaches more and more closely to that giving minimum capital expenditure.

(xii) *Guide to economic velocity*

It is helpful if the designer can start with a rough idea of the range of velocities to include in cost calculations. Clearly, the best guide is previous experience of similar well-engineered systems.

With fluids having properties similar to water, the economic velocity in pump discharge mains of considerable length is likely to be within the range 0·90 – 3·6 m/s (which just covers all cases in the example given in Section Four(d)).

In general, the following velocities are typical:

Boiler feed lines	2·4 – 3·6 m/s
Pump suction lines	0·9 – 1·5 m/s
General Service application	1·5 – 3·0 m/s

It is considered that no attempt should be made here to give guidance for fluids of high viscosity.

(c) Cost of pumps, piping and valves

Approximate costs (late 1966 values) are given (Figs. 4.3 and 4.4) for commonly-used pumps and piping. These prices should be used only as a guide for preliminary design work. Quotations should then be obtained for the equipment shown by the preliminary work to be close to the best economic choice.

The prices given here have been obtained from a limited number of users. In many cases, the user will have access to figures that relate more accurately to the requirements within his organisation.

During the last few years, the annual rise in prices has been substantial. A good indication of current prices in relation to 1966 values can be obtained from the current Monthly Digest of Statistics, published by H.M. Stationery Office.

Materials of construction recommended for a variety of fluids in common use are given in Appendix A.

(i) *Pumps*

Fig. 4.3 gives an approximate indication of prices for commonly-used horizontal centrifugal pumping units with totally enclosed fan-cooled motors.

(ii) *Piping*

Fig. 4.4 gives an indication of piping costs for common materials. The values include for erection at about 6 m working height, and for hangers, but the main supporting steelwork is excluded.

Fig. 4.4 also contains an indication of the cost of lagging for protection against frost, or prevention of condensation of atmospheric water vapour. It is very difficult, however, to formulate general rules for the design of protection required, and the graph given is only a very rough guide.

1. Standard Pump with Cast Iron Case and Impeller, Steel Shaft with Coupling, Guard and Bed Plate
2. As for 1 but with all-Stainless Steel Pump.
3. Unit-Constructed Pump with Cast Iron Case and Impeller, Steel Shaft
4. As for 3 but with all-Stainless Steel Pump
5. Standard Pump with Cast Steel Case, Impeller and Shaft

THE PRICE OF ALL PUMP UNITS INCLUDES T.E.F.C. MOTOR HAVING 2900 R.P.M. SPEED

END SUTION TYPE ⟵|⟶ SPLIT - CASING TYPE

HEAD Metres / Feet	IMP. GAL./MIN. 0–11 (METRES³/HOUR 0–3)	11–44 (3–12)	44–88 (12–24)	88–175 (24–48)	175–350 (48–96)	350–700 (96–192)	700–1400 (192–384)
160 / 520 – 100 / 325							
100 / 325 – 64 / 225	1. £134 2. £291 3. £115 4. £272	1. £302 2. £325 3. £186 4. £276	1. £321 2. £345 3. £186 4. £276	1. £370 2. £1100 5. £550	1. £611 2. £1370 5. £800	1. £1080 2. £1650 5. £1200	1. £2000 2. £4300 5. £3100
64 / 225 – 40 / 130	1. £95 2. £245 3. £84 4. £234	1. £132 2. £260 3. £145 4. £240	1. £132 2. £260 3. £145 4. £240	1. £200 2. £365 3. £198 4. £370	1. £310 2. £1300 5. £570	1. £740 2. £1350 5. £990	1. £1,160 2. £2250 5. £1800
40 / 130 – 25 / 82	1. £35 2. £172 3. £62 4. £115	1. £85 2. £175 3. £66 4. £121	1. £107 2. £220 3. £79 4. £163	1. £160 2. £290 3. £160 4. £277	1. £201 2. £309 3. £201 4. £294	1. £362 2. £620 5. £440	1. £685 2. £1320 5. £850
25 / 82 – 16 / 52	1. £72 2. £142 3. £52 4. £104	1. £80 2. £150 3. £62 4. £114	1. £83 2. £153 3. £64 4. £118	1. £107 2. £258 3. £82 4. £199	1. £182 2. £290 3. £176 4. £277	1. £200 2. £305 3. £186 4. £306	1. £648 2. £1150 5. £730
16 / 52 – 10 / 33	1. £72 2. £142 3. £50 4. £102	1. £75 2. £146 3. £57 4. £109	1. £80 2. £150 3. £62 4. £114	1. £94 2. £172 3. £72 4. £128	1. £111 2. £221 3. £88 4. £149	1. £172 2. £298 3. £161 4. £281	1. £470 2. £1050 5. £600
below 10 / 33	1. £64 2. £142 3. £50 4. £102		1. £80 2. £150 3. £62 4. £114	1. £84 2. £163 3. £64 4. £120	1. £101 2. £211 3. £77 4. £133	1. £152 2. £279 3. £136 4. £256	1. £466 2. £1020 5. £570

Fig. 4.3 Cost of pump units.

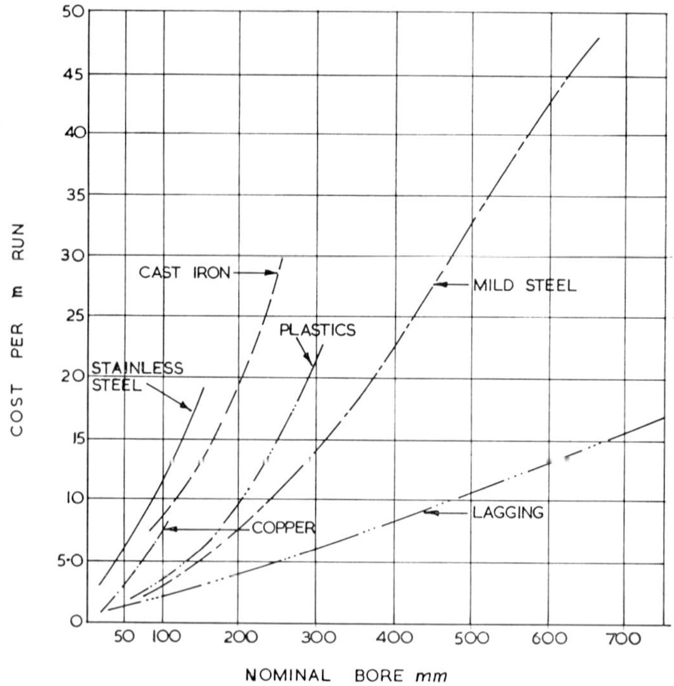

NOTE: 1966 Cost per metre run of pipe, based on 100 m run with one bend, one tee and four flanges supplied and erected to a height of 6 m. All steel pipework and fittings 50 mm and above welded into line; below this size screwed. Plastic pipe coupled by solvent cement joints. Price includes joint rings, nuts and bolts and sufficient pipe hangers fixed to existing steelwork. All materials suitable for working pressure of 7 bar.

Fig. 4.4 Cost of pipework (£).

(iii) *Valves*

No general guide to valve prices is given here because of the large number of types and materials in common use.

(d) Examples of costs

Examples are given below to illustrate the effect of pipe diameter, annual operating time and useful life, on the capital cost and annual pumping cost.

Example 1 is a system with a water pumping duty of 380 m^3/h, a total length of 600 m of mild steel pipe, and no static head. Annual pumping costs are calculated for a range of pipe diameters from 152 to 356 mm (i.e. velocities from 5·79 to 1·06 m/s); for annual working times of 8,760 h (i.e. continuous operation), 2,000 h and 500 h and with a useful life of 10 and 20 years, with 8% interest on capital.

In **Example 2** the same system is treated in exactly the same way, except that a static head of 30 m is superimposed on the friction losses of Example 1.

The costs are shown plotted in Figs. 4.1, 4.5 and 4.6.

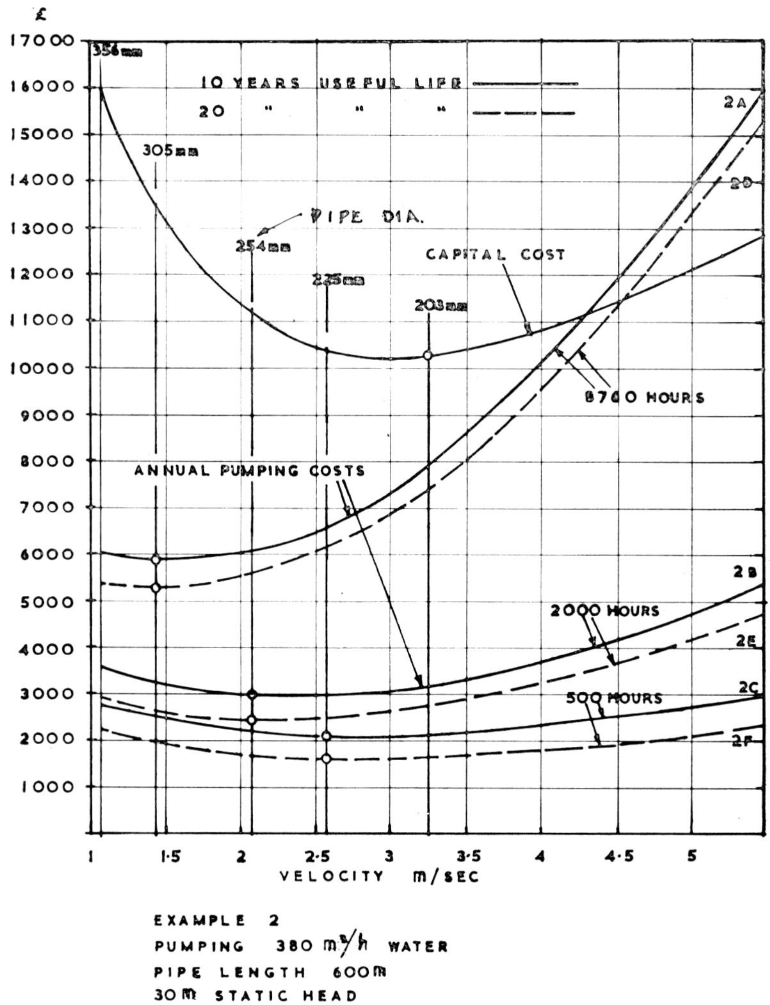

Fig. 4.5 Cost variation: 30 m. static head.

(i) *Remarks on Examples 1 and 2*

1. In these examples the only factor having appreciable effect on the economic choice of pipe size is the annual operating time. The economic pipe sizes are 305 mm for 8,760 h/year, 254 mm for 2,000 h/year, and 203 or 225 mm for 500 h/year. 225 mm is not a standard steel pipe size, but is included to improve the accuracy of the curves. In all cases, the economic choices give velocity within the range 1·2–3·4 m/s, approaching the higher limit as the annual operating time decreases.
2. For continuous operating, the economic pipe size is considerably greater than that giving minimum capital cost. At

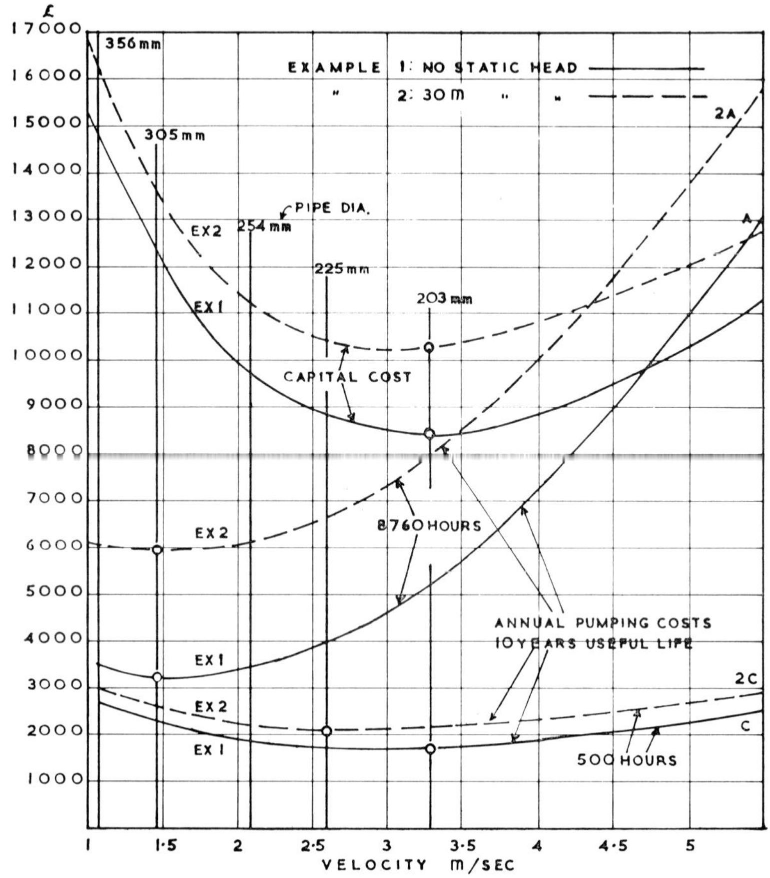

Fig. 4.6 Cost comparison for Examples 1 & 2.

500 h/year these sizes practically coincide, because of the diminished annual energy cost.

3. If the unit energy cost were reduced (as it probably would be with generation by the user) the effect on economic size would be practically the same as given by a proportionate reduction in annual running time with unchanged unit energy cost.
4. It has been assumed that, in all these cases, the electrical installation would have the same cost per kilowatt. In a practical case, there could be a discontinuity in the installation cost/power curve, due to an existing substation having insufficient capacity with the pipe size below a certain value. This could have the effect of increasing the economic pipe size, especially with a system having low annual operating time.
5. In systems required to operate nearly continuously, a standby pump would have to be included. For the sake of simplifying the examples, the provision of one pump only has been assumed in all

cases. Allowance for a standby pump would have a slight tendency to raise the optimum pipe size, because of the progressive rise in the cost of the additional pump with decreasing pipe size.

(ii) EXAMPLE 1.

System to pump 380 m^3/h of water along 600 m length of mild steel pipe with no static head. Losses due to bends, valves, pipe entry and pipe exit are taken as 10 velocity heads in this example (they may alternatively be expressed as equivalent additional lengths in pipe diameters). Calculated for 10 and 20 year's useful life and several values of annual operating time.

Note	(1) Pipe inside diameter (mm)		(2) 152	(3) 203	(4) 225	(5) 254	(6) 305	(7) 356
	Velocity (m/s)		5·82	3·27	2·58	2·09	1·45	1·066
	10 × velocity head (m)		17·06	5·49	3·38	2·22	1·06	0·58
1.	Straight pipe loss (m)		128	31·70	17·67	10·97	4·42	2·13
	Pumping head (m)		145	37	21	13·2	5·49	2·71
2.	Power consumed (kW)		200	51·5	29·2	18·3	7·6	3·75
	Annual cost	8760 h	11700	3010	1710	1070	444	219
	of energy (£)	2000 h	2660	687	389	244	101	50
*	at 1·6d./kWh	500 h	668	172	98	61	25	13
	Capital costs (£)							
3.	Piping		3200	4800	5700	6600	9000	11200
	Valves (5 in number)		220	350	430	520	670	1000
4.	Pump and motor (1 only)		3400	1000	750	600	550	500
	Electrical (£20 per kW)		4000	1030	580	365	150	75
	Civil		1000	1100	1300	1400	1700	2000
	Total	(£)	11820	8280	8760	9485	12070	14775
	Annual capital cost							
	10 years, 8% Total × 0·15		1770	1240	1310	1420	1810	2215
	20 years, 8% Total × 0·102		1206	845	894	968	1232	1510

Notes: 1. From Section One, Fig. 1.16, taking the pipe factor C = 120.
2. Taking pump efficiency × motor efficiency = 0·75.
3. From Fig. 4.3 Class 1.
4. From Fig. 4.4 Mild Steel.
* Old pence

The maintenance costs used in the following cases are, of necessity, rough estimates, and not big enough to have appreciable effect on the economic choice of pipe size. As far as possible, knowledge of actual maintenance costs for comparable systems should be used.

Example 1. Annual Pumping Costs

(1)	(2)	(3)	(4)	(5)	(6)	(7)
Pipe inside diameter (mm)	152	203	225	254	305	356
Case A						
	10 years with 8% interest. 8760 h/year working. Annual capital charge 0·15 × capital cost.					
Annual capital cost	1770	1240	1310	1420	1810	2215
Energy	11700	3010	1710	1070	444	219
Maintenance	260	280	300	310	340	370
Operator wages (3/5 of 1 man)	600	600	600	600	600	600
Total (£)	14330	5130	3920	3400	3194	3404
Case B						
	10 years with 8% interest. 2000 h/year working. Annual capital charge 0·15 × capital cost					
Annual capital cost	1770	1240	1310	1420	1810	2215
Energy	2660	687	389	244	101	50
Maintenance	180	230	250	270	310	350
Operator wages (1/5 of 1 man)	200	200	200	200	200	200
Total (£)	4810	2357	2149	2134	2421	2815
Case C						
	10 years with 8% interest. 500 h/year working. Annual capital charge 0·15 × capital cost.					
Annual capital cost	1770	1240	1310	1420	1810	2215
Energy	668	172	98	61	25	13
Maintenance	180	230	250	270	310	350
Operator wages	60	60	60	60	60	60
Total (£)	2678	1702	1718	1811	2205	2638
Case D						
	20 years with 8% interest. 8760 h/year working. Annual capital charge 0·102 × capital cost.					
Annual capital cost	1206	845	894	968	1232	1510
Energy	11700	3010	1710	1070	444	219
Maintenance	260	280	300	310	340	370
Operator wages	600	600	600	600	600	600
Total (£)	13766	4735	3504	2948	2616	2699

Example 1 *Cont.*

(1)	(2)	(3)	(4)	(5)	(6)	(7)
Pipe inside diameter (mm)	152	203	225	254	305	356
Case E						
	20 years with 8% interest. 2000 h/year working. Annual capital charge 0·102 × capital cost.					
Annual capital cost	1206	845	894	968	1232	1510
Energy	2660	687	389	244	101	50
Maintenance	180	230	250	270	310	350
Operator wages	200	200	200	200	200	200
Total (£)	4246	1962	1733	1682	1843	2110
Case F						
	20 years with 8% interest. 500 h/year working. Annual capital charge 0·102 × capital cost.					
Annual capital cost	1206	845	894	968	1232	1510
Energy	668	172	98	61	25	13
Maintenance	180	230	250	270	310	350
Operator wages	60	60	60	60	60	60
Total (£)	2114	1307	1302	1359	1627	1933

(iii) EXAMPLE 2.

30 m static head. Otherwise the same as Example 1.

(1) Pipe inside diameter (mm.)	(2) 152	(3) 203	(4) 225	(5) 254	(6) 305	(7) 356
From Example 1 (m)	145	37	21	13·2	5·49	2·71
Static head, add (m)	30	30	30	30	30	30
Pumping head (m)	175	67	51	43·2	35·49	32·71
Power consumed (kW.)	243	93·6	71·3	60·4	49·8	46
Annual cost of energy (£) *at 1.6d./kWh — 8760 h	14180	5460	4160	3530	2910	2690
2000 h	3235	1250	950	805	664	613
500 h	810	312	237	201	166	153
Capital costs (£)						
Piping	3200	4800	5700	6600	9000	11200
Valves	220	350	430	520	670	1000
Pump and motor	3800	2000	1400	1200	1000	900
Electrical (£20 per kW)	4860	1870	1430	1210	1000	920
Civil	1000	1100	1300	1400	1700	2000
Total (£)	13080	10120	10260	10930	13370	16020
Annual capital cost						
10 years, 8 % Total × 0·15	1960	1520	1540	1640	2000	2400
20 years, 8 % Total × 0·102	1330	1030	1050	1115	1360	1640
Case 2A — 10 years 8 %. 8760 working annually.						
Annual capital cost	1960	1520	1540	1640	2000	2400
Energy	14180	5460	4160	3530	2910	2690
Maintenance	260	280	300	310	340	370
Operator wages	600	600	600	600	600	600
Total (£)	17000	7860	6600	6080	5850	6060
Case 2B — 10 years 8 %. 2000 working annually.						
Annual capital cost	1960	1520	1540	1640	2000	2400
Energy	3235	1250	950	805	664	613
Maintenance	180	230	250	270	310	350
Operator wages	200	200	200	200	200	200
Total (£)	5575	3200	2940	2915	3174	3563
Case 2C — 10 years 8 %. 500 working annually.						
Annual capital cost	1960	1520	1540	1640	2000	2400
Energy	810	312	237	201	166	153
Maintenance	180	230	250	270	310	350
Operator wages	60	60	60	60	60	60
Total (£)	3010	2122	2087	2171	2536	2963

*Old pence

Example 2 Cont.

(1)	(2)	(3)	(4)	(5)	(6)	(7)
Pipe inside diameter (mm)	152	203	225	254	305	356
Case 2D						
	20 years 8%. 8760 h working annually.					
Annual capital cost	1330	1030	1050	1115	1360	1640
Energy	14180	5460	4160	3530	2910	2690
Maintenance	260	280	300	310	340	370
Operator wages	600	600	600	600	600	600
Total (£)	16370	7370	6110	5555	5210	5300
Case 2E						
	20 years 8%. 2000 h working annually.					
Annual capital cost	1330	1030	1050	1115	1360	1640
Energy	3235	1250	950	805	664	613
Maintenance	180	230	250	270	310	350
Operator wages	200	200	200	200	200	200
Total (£)	4945	2710	2450	2390	2534	2803
Case 2F						
	20 years 8%. 500 h working annually.					
Annual capital cost	1330	1030	1050	1115	1360	1640
Energy	810	312	237	201	166	153
Maintenance	180	230	250	270	310	350
Operator wages	60	60	60	60	60	60
Total (£)	2380	1632	1597	1646	1896	2203

SECTION FIVE

WORKED EXAMPLES

The following examples have been chosen in order to cover as wide a range of use and selection in as few examples as possible. The intention is to show the logic and method clearly, so that this approach can be used in the same general way for different fluids, pipe sizes and working temperatures.

(a) Example one: Variable heads, large quantities, syphonic system

A syphonic condenser cooling system has general characteristics as shown in Fig. 5.1, the top of the condenser being at + 19·5 m, a tidal range being + 2·75 m, and − 2·13 m, relative to the mean tide level at + 0 m. They system discharges over a weir whose sill is at 9·4 m. The condenser is to be supplied with cooling water at 32800 m³/h under mean tide conditions, and under these conditions, the system losses are:

from inlet to condenser input flange	2·13 m
from condenser output flange to weir	1·52 m
losses in condenser, including entry and exit	4·12 m

Pressure is atmospheric at the weir which has a hydraulic characteristic for depth over the weir of

$$h\,(\text{m}) = \frac{Q^{\frac{2}{3}}}{1220} \quad \text{where } Q \text{ is the flow rate } (\text{m}^3/\text{h})$$

Two pumps are to be used in parallel. Select suitable pumps, and investigate the behaviour of the system from the point of view of syphonic conditions with these pumps installed.

(i) *Method*

1. Investigate the system resistance at mean tide condition.
2. Amplify this for max. and min. tides.
3. Derive the mean duty point and specific speed for a suitable pump.
4. Select a pump which on this basis has a suitable performance.
5. From the intersections of this pump's characteristics with the system characteristics, examine its performance at high and low tides.
6. From these duty points, examine the flow conditions and hence the syphonic variations in the condenser.
7. Assuming one pump only running, investigate similarly the maximum syphon in the condenser.
8. At each case considered, check the prime mover rating against pump power requirement.

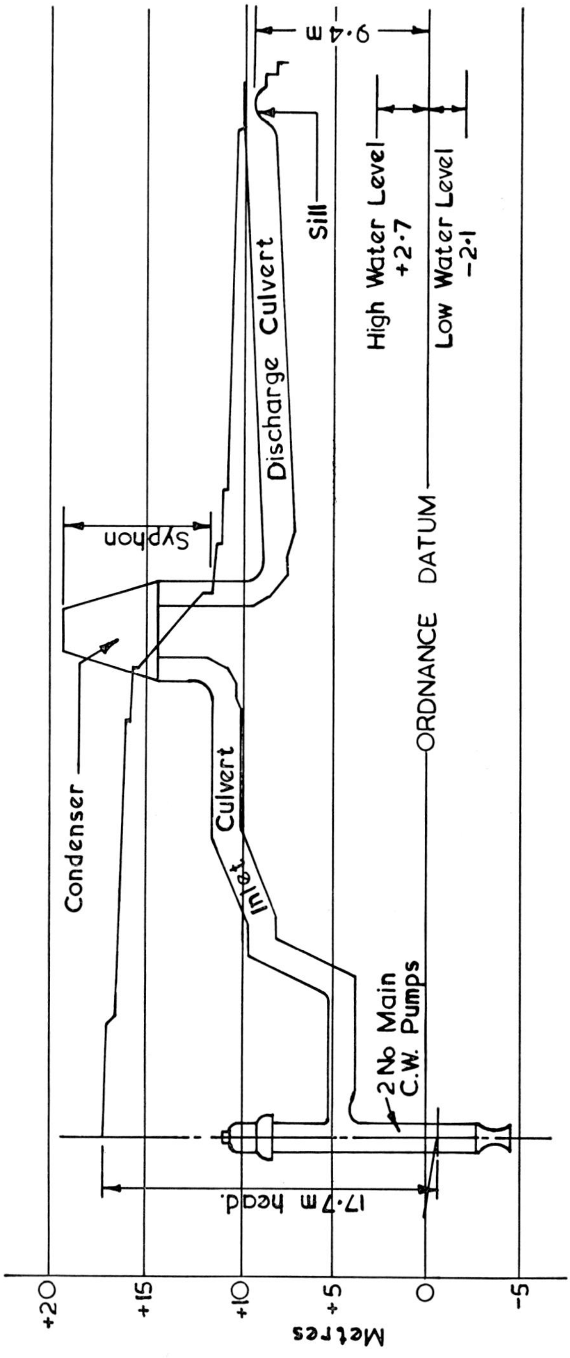

Fig. 5.1 Tidal pumping condenser system.

(ii) *Calculation and logic*

1. *System resistance*

The resistance varies with v^2 and since the sectional area is constant and $Q = Av$ the resistance varies with Q^2

Total frictional losses at duty point $= 7{\cdot}77$ m

Static lift to sill of discharge weir $= 9{\cdot}4$ m

The system characteristic at mean tide level can now be drawn (marked (1) on Fig. 5.2) on the basis that the curve is parabolic and is given by

$$Q = \sqrt{4a\,(H - a)}$$

where a = static lift intercept on vertical (H) axis. There is an additional lift equal to the height of water over the weir, usually referred to as the 'nappe'. This is given by

$$h = \frac{Q^{\frac{2}{3}}}{1220}$$

and varies with the flow of water.

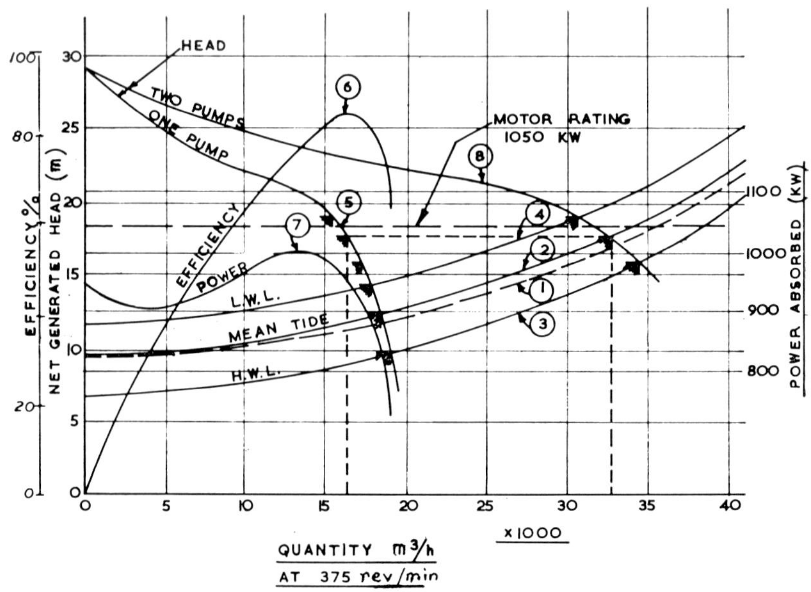

Fig. 5.2 Syphonic cooling system with variable heads.

At 32800 m³/h., the depth is 0·84 m. Similarly, depths can be calculated for other flows. The true overall curve is, however, inclusive of added static lift, as the depth of flow over the weir increases with quantity, and to the system resistance curve (1) the weir depths are added, thus producing curve (2).

2. *Tidal variations*

The tidal variations are represented by shifts of curve 2 within the extremes of tides by an equal vertical ordinate at all points to the tide variation, i.e.: +2·75 m and −2·13, thus giving curves 3 and 4, for high tide and low tide respectively.

3. The mean duty point is now seen to be at 32, 800 m³/h with 17·7 m head; i.e. 16, 400 m³/h per pump at 17·7 m head, which at 375 r.p.m. gives a specific speed of 4,380 and indicates a mixed flow type.
4 A pump proposed for the system has characteristics as shown in curve 5 ($Q-H$ curve), curve 6 (efficiency) and curve 7 (power). With two such pumps in parallel, curve 8 is obtained by adding horizontal quantities for any fixed value of head.
5. From the two pumps in parallel characteristic, it will be seen that the following are the likely duty points at which the pumps will operate:

Mean tide	32, 800 m³/h	17·7 m head
High tide	34, 400 m³/h	15·85 m head
Low tide	30, 300 m³/h	18·8 m head

6. Having established these figures, the mean tide syphonic variations in the condenser can be estimated as follows:

Basic height of sill at weir	9·40 m
Height of nappe at mean flow	0·84 m
Losses on discharge side of condenser at mean flow 32, 800 m³/h	1·52 m
Total	11·76 m
Height of condenser top	19·50 m
Syphon developed	7·74 m

Similarly, since losses vary with the square of the quantity pumped, the discharge losses at 30, 300 m³/h (low tide) are 1·32 m and also at this flow the weir nappe is changed to 0·79 m These figures substituted in the above calculation give a syphon of 8·03 m.

Further substitution for the high tide figures give a syphon of 7·60 m.

Not too much notice should be paid to the figures behind the decimal point in the final total of calculations of this sort; they are included for the sake of arithmetical clarity. Again, hydraulic gradients covering the cases above are drawn on Fig. 5.2 marked 'mean', 'high' and 'low' after calculation of losses through the system on the same basis as above, i.e. losses vary before and through the condenser as the square of the velocity and hence the square of the quantity. It has been assumed, for the sake of this exercise, that the losses are even through the parts of the system concerned, but in practice sharp losses will occur at bends, tees, etc., and the correct shape is shown in Fig. 5.2.

7. If one pump only is operating, the condition of operating will be as follows:

Mean tide	18, 300 m^3/h at 12·2 m head
Low tide	17, 700 m^3/h at 14·3 m head
High tide	19, 100 m^3/h at 9·75 m head

The maximum syphon will occur in the condenser when the flow is a minimum, i.e. when losses beyond the condenser are a minimum.

The corresponding figures then become as before

Height of sill	9·40 m
Nappe at min. flow	0·55 m
Losses on condenser discharge side	0·44 m
Total	10·39 m
Height of condenser top	19·50 m
Max. syphon	9·11 m

8. From examination of the pump power characteristic curve, it will be seen that the mean power requirement is 955 kW. Variations for high and low water are 925 kW and 980 kW respectively. These are within 10% of the mean, and should not cause trouble in the prime mover, if ordered for mean duty, and a 10% continuous overload.

It is possible that many pumps would meet the duties shown above, and comparison of pump/motor units should proceed at tender stage to cover the economics of running against capital cost and general suitability. In this design, the pump should be capable of priming the system and in this case the maximum head of the pump characteristic at a low flow must equal the static lift to the top of the condenser from a low tide condition – in this case 19·5 + 2·13 = 21·63 m. It will be noted that the power curve is less than its maximum at low flows and at the closed valve condition. No starting problems therefore arise.

However, it will be seen that an intermediate point on the pump power curve rises to 1000 kW, and the system could be held at this point due to a semi-closed valve. At a 955 kW rating, this is still marginally within the 10% overload. In practice, the nearest manufacturer's size of motor will be used and in this case the motor rating concerned is 1050 kW, which amply covers all working conditions.

(b) Example two: simplified boiler feed system

The following is an example dealing with a simplified boiler feed system. (i) illustrates the method of determining whether flow in the system is laminar or turbulent, (ii) indicates the method of finding the total frictional head loss in the system, while (iii) deals with the method for calculating the head to be generated by the boiler feed pump. It is sometimes necessary for the pump duty to be changed, and this is achieved by the method shown in (iv), whilst (v) indicates the method adopted for determining the suction head that must be provided for the pump to ensure cavitation-free operation.

(i) Water at 150°C. is flowing along a 355·5 mm diameter pipe at 408·23 kg/s. The density of water at 150°C. is 917·43 kg/m³. Determine whether the flow in the pipe is laminar or turbulent. Dynamic viscosity of water at 150°C. is $18{\cdot}6\times10^{-5}$ Ns/m².

$$a=\frac{\pi D^2}{4}=\frac{\pi}{4}(0{\cdot}3555)^2=0{\cdot}0992\ \text{m}^2$$

$$Q=408{\cdot}23\ \text{kg/s.}$$

$$=\frac{408{\cdot}23}{917{\cdot}43}=0{\cdot}445\ \text{m}^3/\text{s.}$$

$$V=\frac{Q}{a}=\frac{0{\cdot}445}{0{\cdot}0992}=4{\cdot}485\ \text{m/s}$$

$$R_e=\frac{vDw}{\mu}=\frac{4{\cdot}485\times0{\cdot}3555\times917{\cdot}43}{18{\cdot}6\times10^{-5}}$$

$$=7{\cdot}861\times10^6$$

Since the Reynolds Number is above 3000, the flow is clearly turbulent. Temperature of the fluid concerned has a considerable effect on Reynolds Number. In the above example, if the water were at 16°C., R_e would be 131,000.

(ii) Calculate the total frictional head loss in the following system (Fig. 5.3) including the valves at exit and entry.

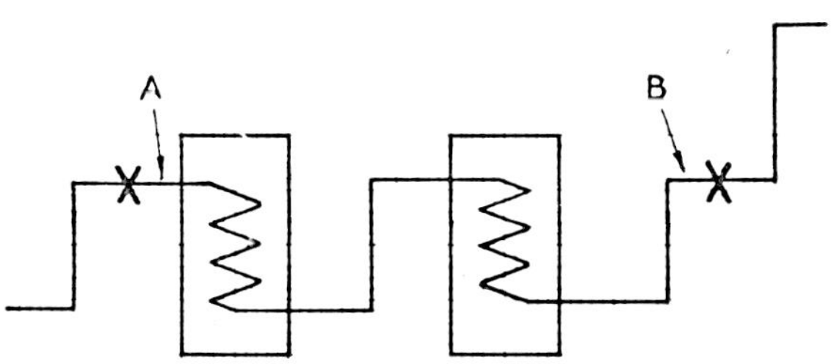

Fig. 5.3 Boiler feed system.

Flow = 408·23 kg/s.
Pipe diameter = 355·5 mm
Head loss between A and B at 408·23 kg/s = 23 m
Total length of pipe, excluding that between A & B = 152 m.

Consider the gate valves to be fully open, and all bends to be long sweep elbows. Find the equivalent length of straight pipe to one gate valve fully open, by using the nomogram of Fig. 1·17

This gives an equivalent pipe length of 2·438 m.
Equivalent length for 4 valves = 9·754 m.
Similarly, equivalent length for 4 long sweep elbows from the same nomogram = 4 × 7·01 = 28·04 m.

Thus total length of pipe to be considered

$$= 152 + 9{\cdot}754 + 28{\cdot}04$$

$$= 189{\cdot}794 \text{ m}$$

Velocity in pipe from (i) $= 4{\cdot}485$ m/s.

The friction factor, assuming the pipe to be smooth, can be found from the nomogram, Fig. 1–10

$$\text{Kinematic viscosity} = \frac{\text{dynamic viscosity}}{\text{specific weight}}$$

$$= \frac{18{\cdot}6 \times 10^{-5}}{917{\cdot}43}$$

$$= 2{\cdot}027 \times 10^{-7} \text{ m}^2/\text{s}$$

Since this value of viscosity is beyond the range of the right-hand scale of viscosity in Fig. 1.10 a graph of friction factor against viscosity for a pipe size of 355·5 mm and velocity 4·485 m/s has been drawn, using values taken from the nomogram (Fig. 5.4).

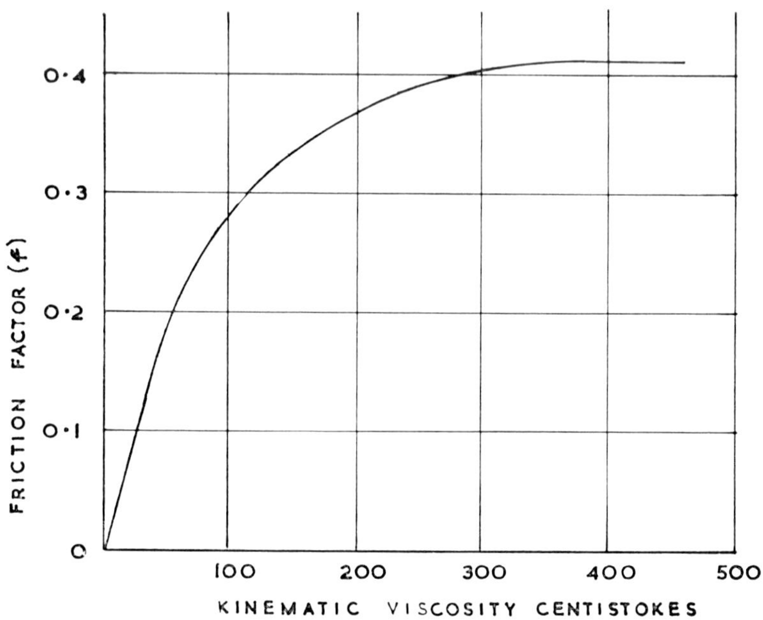

Fig. 5.4 Friction factor – viscosity.

It can be seen from this curve that the friction factor becomes very small at high values of viscosity, but to facilitate the use of the nomogram, the friction factor for this example will be taken as 0·01.

The head loss per metre run, found from nomogram Fig. 1.12, is 0·028 m/m run.

The pressure drop, obtained from the same nomogram $= 2{\cdot}7 \times 10^{-3}$ bar/m run.

∴ Head drop over 189·794 m = 189·794 × 0·028 = 5·313 m
Hence, total frictional head loss in system = 5·313 + 23 = 28·313 m

(iii) Consider now the system extended as Fig. 5.5 with a feed pump provided. It is necessary to find the head to be generated by the pump.
Frictional loss in suction pipe at 408·23 kg/s = 2·438 m

(1) Nett static pressure:

Steam drum pressure	= + 172·2 bar
Feed pump centre-line to drum water level (= 38 m) at 150°C.	= + 3·423 bar
Deaerator steam pressure	= − 4·477 bar
Deaerator water level to feed pump centre-line (= 30·5 m)	= − 2·613 bar
∴ Nett static pressure	= 168·533 bar

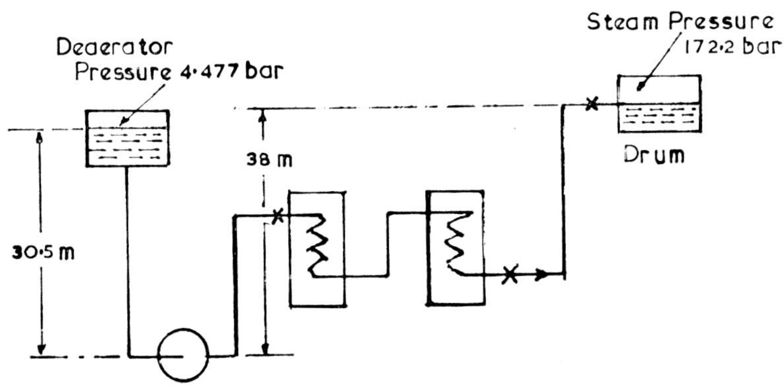

Fig. 5.5. Boiler feed system with feed pump installed

(2) Total frictional loss at rated flow = loss in suction pipe + loss in discharge pipework (from (ii))
= 2·438 + 28·313
= 30·751 m
= 2·756 bar

System resistance = nett static head + total friction loss
= 168·533 + 2·756
= 171·289 bar

Taking a design margin of 3%, system resistance at rated duty
= 171·289 × 1·03
= 176·40 bar
= 1963 m

The required duty of the pump is therefore 1963 m at 408·23 kg/s.

The power required at this duty (assuming a pump efficiency of 85%) is given by:

$$\frac{9{\cdot}807 \times Q\ H}{\text{pump efficiency}}$$

$$= \frac{9{\cdot}807 \times 408 \times 1963}{0{\cdot}85 \times 1000} = 9260\ \text{kW}$$

where $Q = \text{kg/s}$, $H = \text{m}$.

A pump of this duty will be a multi-stage centrifugal design, with possibly 4 stages.

$$\therefore\ \text{Head per stage} = \frac{1963}{4}$$

$$= 490{\cdot}75\ \text{m}$$

Assume the pump speed (N) to be 5100 r.p.m.

$$\text{Specific speed of each stage} = \frac{N\sqrt{Q}}{H^{\frac{3}{4}}}$$

where $Q = \text{m}^3/\text{s}$, $H = \text{m}$

$$408{\cdot}23\ \text{kg/s} = \frac{408{\cdot}23}{917{\cdot}43} = 0{\cdot}455\ \text{m}^3/\text{s}$$

$$\text{Specific speed of each stage} = \frac{5100\ \sqrt{0{\cdot}445}}{(490{\cdot}75)^{\frac{3}{4}}}$$

$$= \frac{5100 \times 0{\cdot}664}{32{\cdot}97}$$

$$= 32{\cdot}4$$

(iv) Consider now that the system is modified slightly, so that the generated head of the pump becomes 1940, that is, 485 m per stage. This reduction in head can be achieved by reducing the diameter of the impeller.

Let H_1, Q_1 and D_1 be the head, flow, and diameter of the original pump impeller, and H_2, Q_2 and D_2 the head, flow, and diameter of the impeller for the new duty.

$H_1 = 1963$ m $\qquad D_1 = 400$ mm

$H_2 = 1940$ m $\qquad D_2 =$ unknown.

$$\frac{H_1}{H_2} = \left[\frac{D_1}{D_2}\right]^2$$

$$D_2 = D_1\sqrt{\frac{H_2}{H_1}}$$

$$= 400\sqrt{\frac{1940}{1963}}$$

$$= \underline{398{\cdot}4\ \text{mm}}$$

Reducing the diameter of each impeller from 400 mm to 398·4 mm will give a reduction in total generated head from 1963 to 1940 m. However, it will also result in a slight decrease of flow as follows:

$$\frac{Q_1}{Q_2}=\frac{D_1}{D_2}$$

$$=\frac{408{\cdot}23}{Q2}=\frac{400}{395{\cdot}3}$$

$$Q_2=\frac{408{\cdot}23\times 398{\cdot}4}{400}$$

New flow rate = 406 kg/s.

(v) To find the NPSH (net positive suction head) available at rated flow rate.

NPSH available $=h_a+h_s-h_f-h_{vp}$

where h_a = pressure in deaerator
h_s = static head.
h_f = friction loss in suction pipe
h_{vp} = vapour pressure of fluid

Since the deaerator pressure is equal to the vapour pressure in this example, the equation becomes:

NPSH available $=h_s-h_f$
$=30{\cdot}5-2{\cdot}438$
$=28{\cdot}062$ m

It is important, therefore, to ensure that the pump is designed to have a required NPSH value below 28·062 m. In a boiler feed pump system, it is usual to arrange for the available NPSH to be between 70 and 100% greater than the required NPSH, to allow for transient operating conditions.

It is generally not possible to design a pump of this duty to operate satisfactorily on the suction head available. It is thus usual to have two pumps in series. The first pump has a generated head of approximately 150 m, which will supply sufficient suction head to the main pump, which will generate the remainder of the head.

The characteristic curves for the two pumps are shown in Fig. 5.6.

(c) Example three: Pumping kerosene

Kerosene at 15°C. and specific gravity (S.G.) 0·815 is to be pumped at 1·52 m/s through a 152 mm diameter steel main 1520 m long. The system has one gate control valve, three bends and a swing check valve, and is subjected to a static head of 33 m. Estimate the operating point of a suitable pump and its power requirements.

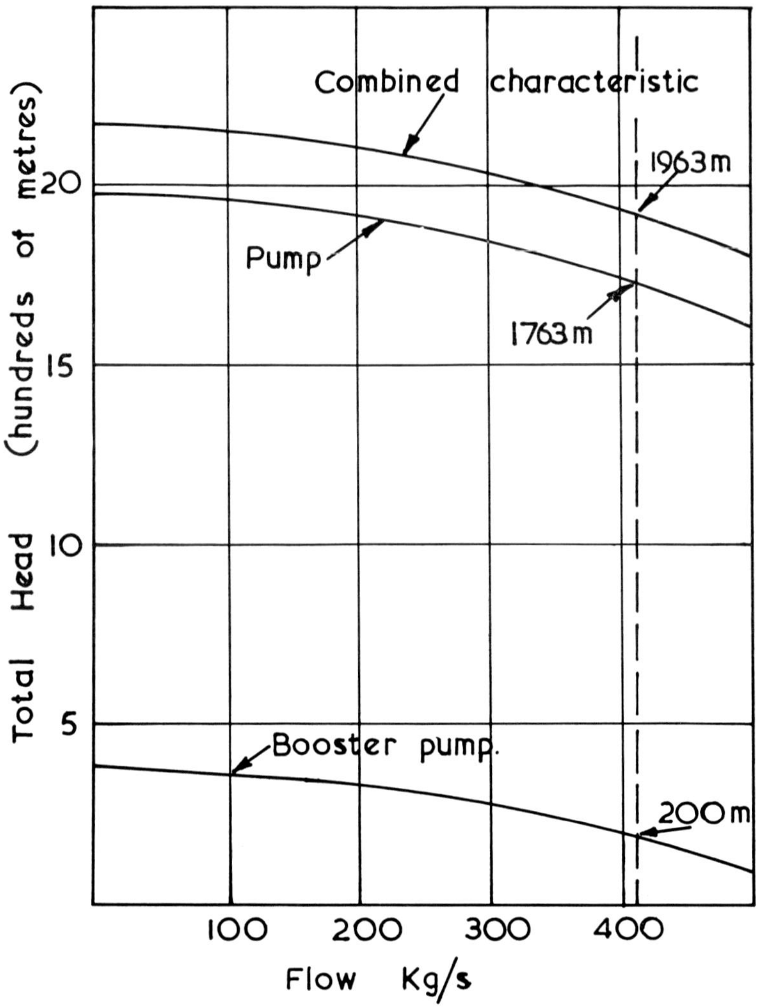

Fig. 5.6 Series pumping – booster and main boiler feed pump

If the system capacity were to be increased by 10%, estimate the increase in pump impeller diameter at the new duty point, and the new power requirement.

(i) *Method*

1. Determine the viscosity condition.
2. Determine the Reynolds Number and flow conditions.
3. Determine appropriate roughness factors, and hence the friction factor.
4. Calculate the head losses.
5. Calculate the duty point and power requirement.

6. Calculate revised h_f valves at new system load, and new duty point.
7. Estimate impeller diameter increase and revised power requirement.

(ii) *Calculation:*

1. *Viscosity:*

Absolute viscosity of kerosene at 15°C. $= 2$ centipoise $= 2 \times 10^{-3}$ Ns/m²

Kinematic viscosity of kerosene at 15°C. $= \frac{2}{SG} = \frac{2}{0{\cdot}815}$
$= 2{\cdot}46$ centistokes
$= 2{\cdot}46 \times 10^{-6}$ m²/s

Using Fig. 1.9 for a specific gravity of 0·815 and 2 centipoise gives 2·46 centistokes.

2. Reynolds Number

$$R_e = \frac{vD}{\gamma} = \frac{1{\cdot}52 \times 0{\cdot}152}{2{\cdot}46} \times 10^6$$
$$= 9{\cdot}4 \times 10^4$$

R_e is greater than 3000 and therefore flow is turbulent.

3. Roughness factors:

From Table 1, the k value for steel pipe uncoated is 0·00003

$$\text{Relative roughness} = \frac{k}{D} = \frac{0{\cdot}00003}{0{\cdot}152} = 0{\cdot}0002$$

From Fig. 1.15 for $\frac{k}{D} = 0{\cdot}0002$ and $R_e = 94000$

friction factor $= 0{\cdot}02$

Note that for a smooth pipe at this value of R_e, the friction factor is 0·018.

Alternatively from Fig. 1.10 connecting 2·46 centistokes kinematic viscosity with 0·152 m dia. of pipe, find the reference line point. Through the same point draw a line from the 1·52 m/s velocity scale, which gives R_e about 95,000. From this value of R_e draw a line through the Turbulent Flow reference centre, to give a smooth pipe friction factor of 0·018 as before, Multiplying this by the relevant factor in Table 1.1 gives $f = 0{\cdot}018 \times 1{\cdot}1 = 0{\cdot}02$.

4. *Head loss*

Equivalent lengths of fittings are given in Fig. 1.17 as follows, reading from the 0·152 m diameter ordinate on the right.

Check valve	= 12 m
Gate valve	= 1·2 m
Bends: 3 at 5 m	= 15·0 m
Total fittings	= 28·2 m

Pipeline $= 1520{\cdot}0$ m

Total system $= 1550$ m say

$$h_f = \frac{fLv^2}{2gD} = \frac{0{\cdot}02 \times 1550 \times (1{\cdot}52)^2}{2 \times 9{\cdot}81 \times 0{\cdot}152}$$

$= 24$ m say

Static head $= 33$ m

Total head, dynamic $= 57$ m

$$\text{Flow quantity} = \frac{1{\cdot}52 \times \pi \times (0{\cdot}152)^2}{4} \times 60 \times 60$$

$= 100\ \text{m}^3/\text{h}$

5. The duty point is therefore 57 m head at 100 m^3/h.
6. Revised duty:

h_f varies directly with v^2 since a 100 % change in flow rate will not materially affect R_e and f. A check on this is as follows:

Change of velocity required $= 10$ %, since pipe area is constant

Since v is increased by 10 %, with no change in $\frac{D}{\gamma}$, the R_e value is increased by 10 %

New $R_e = 1{\cdot}1 \times 95000 = 104{,}500$

$\frac{k}{D}$ remains as before $= 0{\cdot}0002$

From Fig. 1.15, $f = {\cdot}0195$

Comparing this with the previous value of ·02, the difference in terms of practical hydraulics is negligible.

It will be assumed, then, that h_f varies directly with v^2

$$\text{New } h_f = h_{f1}\left[\frac{v}{v_1}\right]^2 = 24 \times 1{\cdot}1^2 = 29 \text{ m}$$

New total head $= 29 + 33 = 62$ m

New flow rate $= 100 \times 1{\cdot}1 = 110\ \text{m}^3/\text{h}$ and the new duty point will be 62 m at 110 m^3/h.

7. Change of impeller diameter

An approximate curve is drawn (Fig. 5.7) through the duty point representative of the pump characteristic concerned, and the method outlined in Section Three (f) applied:

chosen reference point 80 m^3/h

$$\text{corresponding head} = 62\left[\frac{80}{110}\right]^2 = 45 \text{ m}$$

The reference point and the new duty point are joined and cut the existing pump characteristic at 101 m^3/h flow rate.

The likely increase in impeller diameter is then

$\frac{110}{101} = 1{\cdot}09$, that is, a change of 9 %.

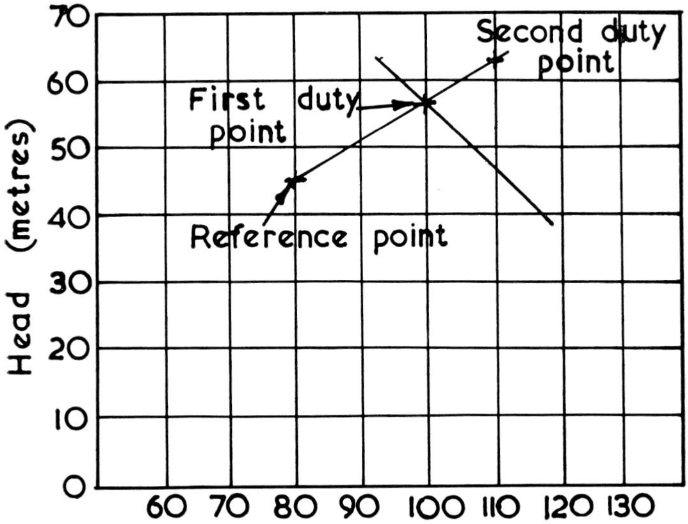

Fig. 5.7 Example: change of impeller diameter.

(Q measured in m^3/h)

SECTION SIX

DESIGN AND USAGE

(a) Commercial equipment available

Pump usage has become so general in industry that a detailed survey of available types and variations is beyond the scope of this handbook.

Such information is best obtained from makers' catalogues, and this section is necessarily confined to guidance on the main categories of usage, with comments on the general intention of any particular design.

Impellers of particular types are shown in Figs. 6.5 to 6.8.

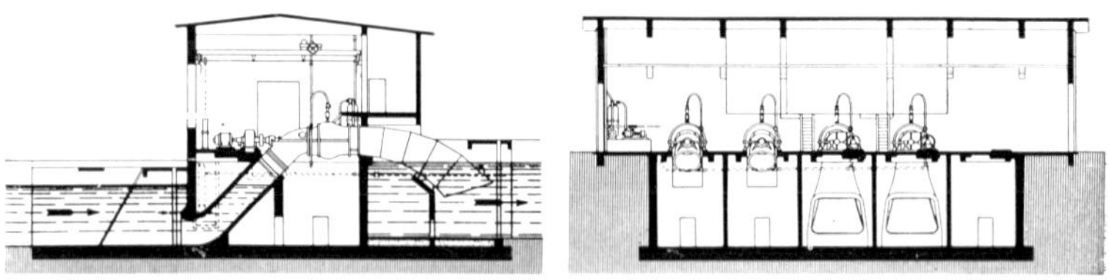

Horizontal Arrangement

Balamoan Irrigation Plant on the Nile Delta. The mixed-flow pumps are fitted above flood level at the highest point of the pipes which connect the suction and delivery channels in the manner of a siphon. Any reflux of water when the pumps are stopped can be prevented by opening an air admission valve.

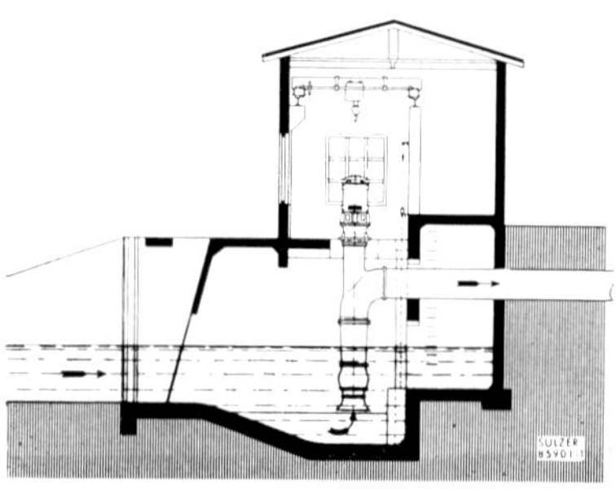

Vertical Arrangemen

Sarvaz Drainage Plant, Switzerland. Three vertical mixed-flow units together raise 28644 m^3/h against 4·654 m. The territory drained by this plant totals 300 ha at high water.

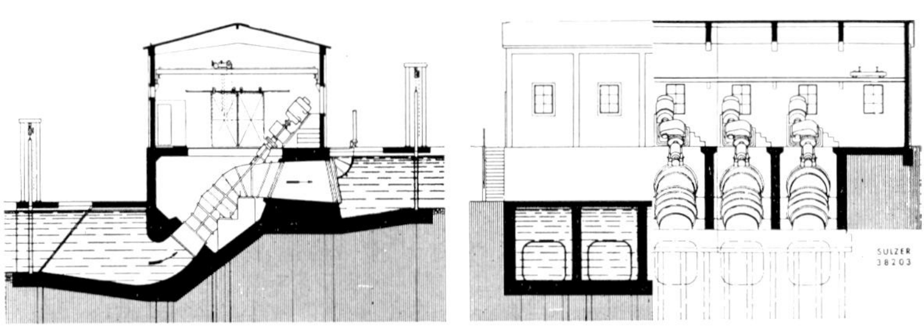

Inclined Arrangement

Drainage plant with three inclined mixed-flow pumps. Electric motor drive through gearing. As the impellers are immersed in the water, the pump can be started without previous priming. Automatic flap valves prevent reflux.

Fig. 6.1 Horizontal, vertical and inclined pump arrangements.

(i) *Axial flow types*

In general, the axial flow pump is largely confined to main drainage and low head condenser cooling duties. Fig. 2.1 shows details typical of this type of pump and Fig. 6·1 gives a selection of normal use in which, as can be seen, it is usually mounted in a wet well with the impeller immersed.

It has three major sections to its construction: the bell-mouth entry to smooth inlet flow conditions, the axial type of impeller mounted in a pipe-type casing, and a discharge section angled to suit the installation. The drive is effected through the angled discharge casing from a motor usually mounted integral with (or bolted to) the casing.

Axial flow pumps can be very large, and may have features specially aimed at particular variations of duty. Head conditions may not be constant in the duties concerned, and some models provide means whereby the angle of attack of the impeller can be changed, whilst running, to meet such changes.

The pumps are essentially simple in construction and easy to maintain. Particular attention should be paid to the adequacy of mounting and the nature of the thrust bearings. It should also be noted that the motor must be above maximum water level and this is effected in two ways:

1. By extending the delivery tube so that the necessary height is achieved as in Fig. 6.2(a) or, since this is not always convenient,
2. by forming an 'occasional' dry well as shown below,

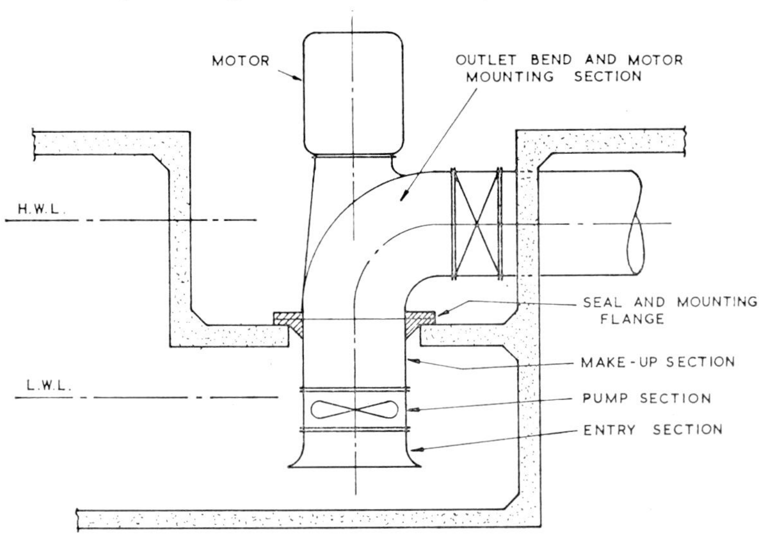

(a) SEMI-DRY WELL

Fig. 6.2 Mounting of axial/mixed flow vertical pumps.

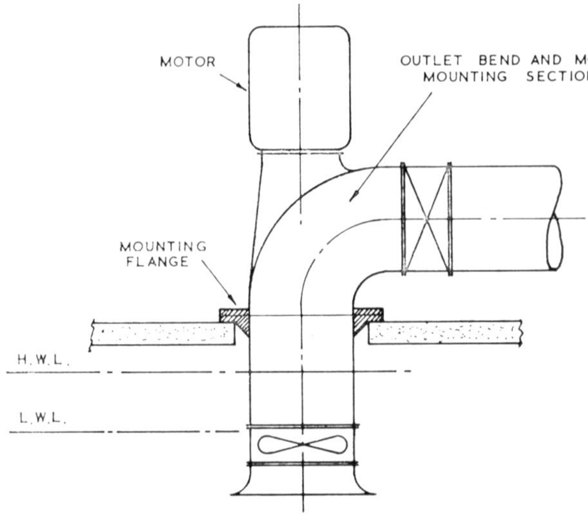

Fig. 6.2 Mounting of axial/mixed flow vertical pumps.

where the motor/discharge-bend mounting is sealed to the floor to prevent flooding at high water level, and may be dismantled in the dry at low water levels.

Attention should always be paid to the headroom requirements of vertical pumps during dismantling or assembly. By making the casing and shaft in sections, the headroom can be reduced considerably, as also is the case by using the technique outlined in 2 above.

(ii) *Mixed flow pumps*

Horizontal or vertical types are available, being recognised from the axial type by the more bulbous pump casing (Fig. 6.3). As heads increase, volute casings are necessary (Fig. 6.4), and at still higher heads, the mixed flow pump merges naturally into the centrifugal range. The mixed flow pump is also available in multi-stage form.

Various arguments exist with respect to the use of horizontal or vertical units. These usually centre around the lesser space and simpler building requirement of the vertical unit on the one hand, whilst others advocate the split casing horizontal pump for ease of maintenance. Differences in the nature of bearings form another seat of argument, some favouring the journals of the horizontal pump as against the thrust bearings of the vertical models. With modern bearing techniques such arguments have receded, and the only logical way of selection should be that of overall cost of the installation, inclusive of its civil engineering components.

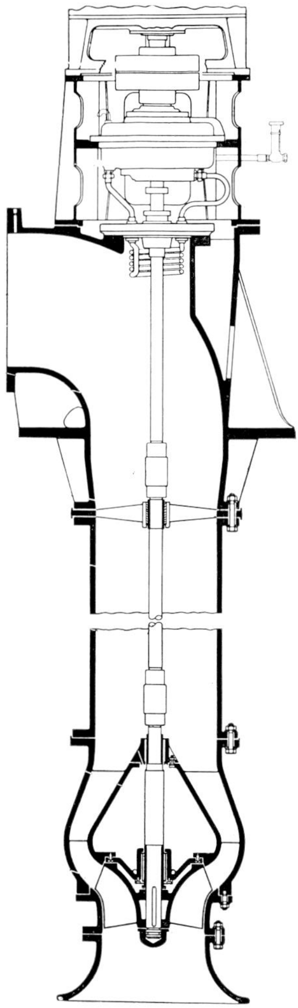

Fig. 6.3 Mixed flow diffuser pump vertically suspended.

(iii) *Centrifugal pumps*

By reason of variety of industrial application, the greatest variety of pumps, by far, is in this range. Broadly, classification is by the nature of the impeller, whether it is single or double entry, or in multi-stage form. Single entry impellers are for the lesser capacity medium head range, double entry covering the larger capacities, and multi-stage pumps covering the very high head ranges. Further, particularly in the single entry type, there are many impeller variations to meet special applications. High efficiency pumps are designed in a manner where the water path through the impeller is under maximum control, hence the enclosed type of impeller. For lesser efficiencies and lower costs,

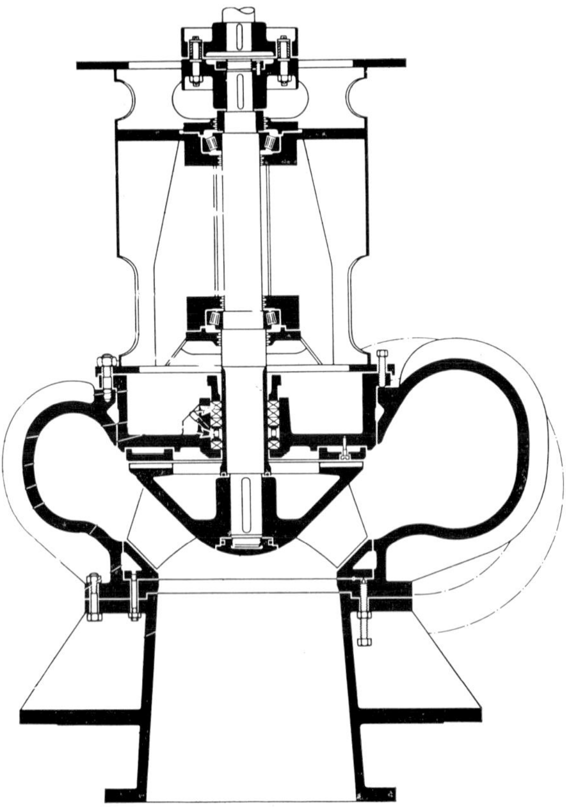

Fig. 6.4 Mixed flow volute pump vertical spindle.

impellers of the open or semi-closed types are available. Certain impellers are specially designed for handling semi-solids or abrasive materials and are said to be of the 'non-clogging' type. One particular feature of impeller design should be noted: that of pressure balancing. Fig. 6.9 shows an impeller in a casing with, at (a), a single wearing ring and at (b) two wearing rings and a shouldered impeller. It will be seen that in (a), pressures on the impeller are unbalanced due to the larger back area in the higher pressure zone, whilst with the shouldered design (b), the back of the impeller is bored through, giving an area of equal pressure to that on the suction side, with the second wearing ring at the shoulder forming a pressure change orifice. Double entry impellers are, of course, balanced by nature of their entry conditions. High efficiency centrifugal impellers are invariably of the doubly enclosed type, but it should be noted that with age (or wear), the efficiency will drop quicker than that of the simpler impeller, which has a lower but (in terms of age) more stable efficiency. The reason for this is wear at the rings, and a greater surface area for wear on the impeller itself. Wearing rings

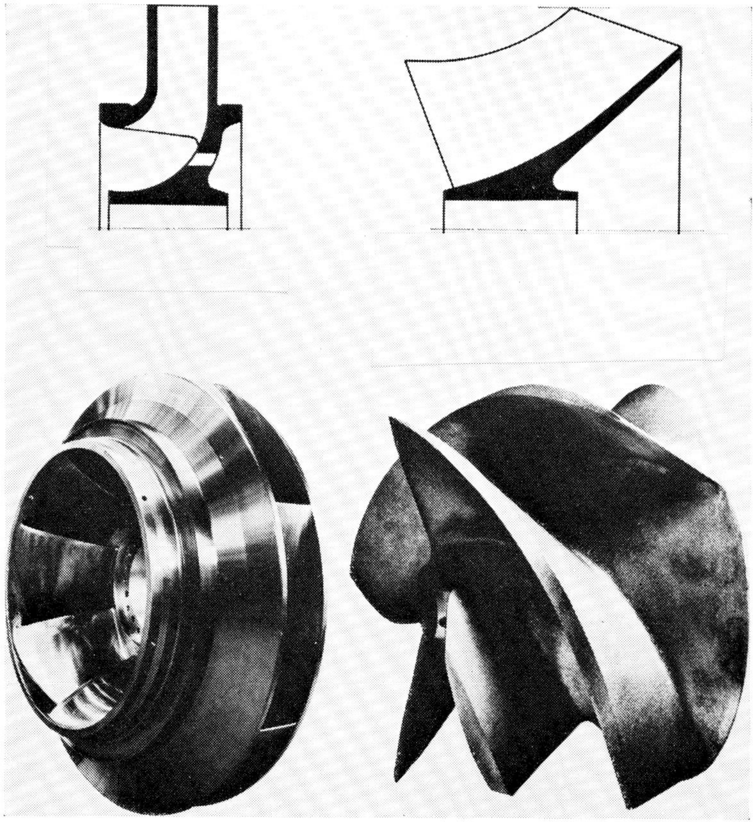

Fig. 6.5 Mixed flow impellers. (Left) A Francis vane impeller. (Right) An open mixed flow impeller.

Fig. 6.6 Centrifugal impellers (a) A shrouded centrifugal impeller.

Fig. 6.6 Centrifugal impellers (b) A semi-shrouded impeller.

Fig. 6.7 A three-bladed axial flow impeller.

Fig. 6.8 Non-clogging impellers.

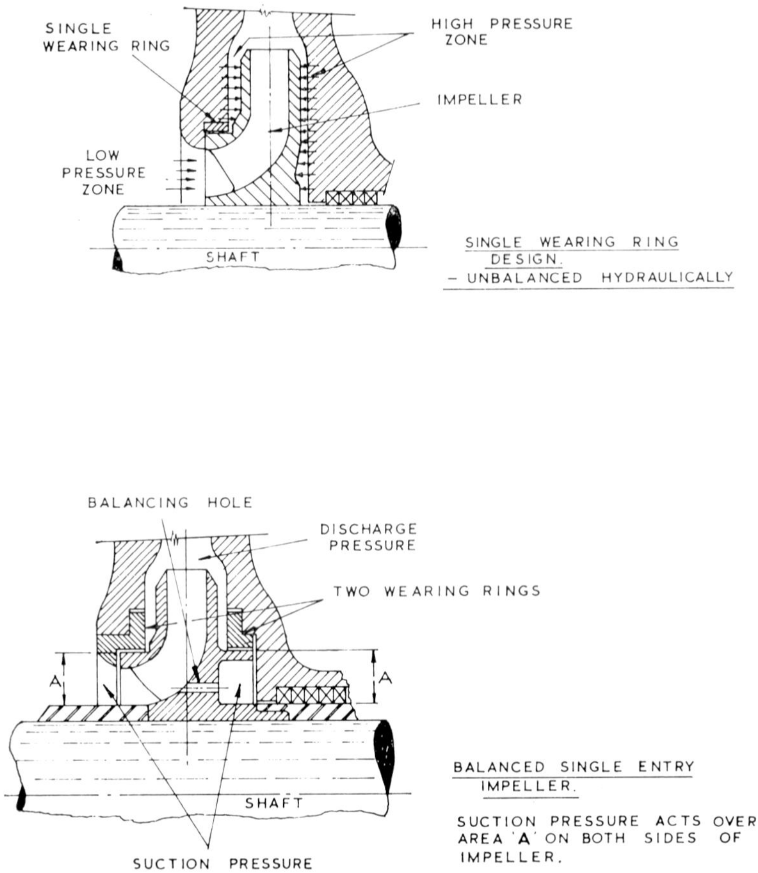

Fig. 6.9 Unbalanced and balanced impellers.

should, of course, be closely inspected at maintenance periods and renewed where necessary.

As a division of pump types as opposed to impeller types, the following examples serve to illustrate the range of availability. Various terms are used generally to describe any particular type, and the terms used in the following sampling are indicative of the use and type, rather than definitive:

1. *General purpose overhung pump*

   This is a simple robust pump for general use in fluids compatible to its materials of manufacture. In its basic form it may have an unbalanced impeller, and in general, the pump is overhung on a mounting bracket (Fig. 6.10), with end-suction condition from a straight inlet branch through which the impeller can be inspected. Further dismantling of the end cover enables the impeller to be withdrawn and sealing rings examined for wear. Most centrifugal impeller types may be seen housed in

---

Fig. 6.8 Non-clogging impellers
Top – Typical double-shrouded sewage impeller.
Bottom – Three-bladed unshrouded, non-clogging impeller.

this form of pump. The bearings, glands and casing are of the simplest construction possible, commensurate with proper operation.

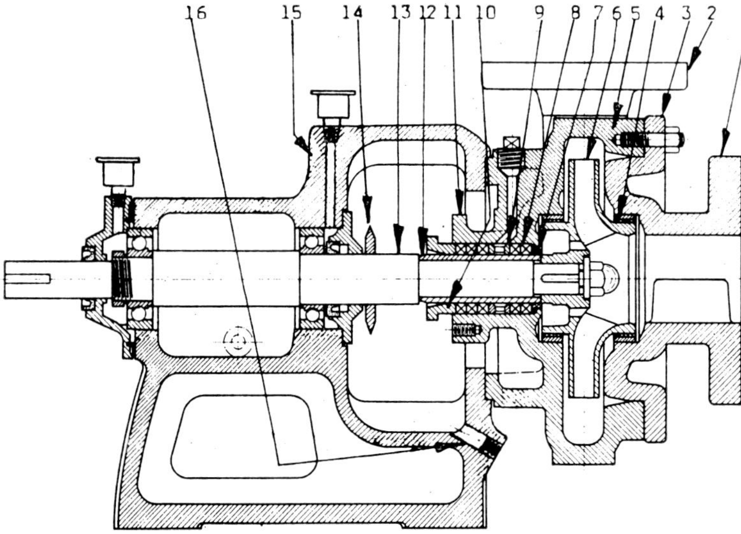

1 Suction branch
2 Discharge branch
3 End cover
4 Case and impeller wearing rings
5 Casing
6 Impeller
7 Neck ring or bush
8 Packing
9 Lantern ring
10 Gland
11 Stuffing box
12 Shaft sleeve
13 Shaft
14 Liquid thrower ring
15 Frame
16 Gland well drain connection

Fig. 6.10 Typical end suction, frame-mounted centrifugal pump.

2. *Heavy duty process pumps*

These appear in a variety of forms, and in materials suitable for the process concerned.

A variety in which the pump body can be removed without dismantling the casing and pipework is shown in Fig. 6.11, the end casing being usually pedestal supported and a further support being located under the shaft bearing block. For removal, the pump is uncoupled from the motor at a spacer coupling and the bearing housing with the complete rotating assembly removed from the impeller housing, which enables all parts normally subject to wear (except the case wearing rings), to be removed as a unit and attended to at workshop facilities. Stand by pump units are quickly inserted and outage-time cut to a minimum.

Such pumps are of robust construction and cooling jackets are provided for the stuffing box and bearing brackets.

Heavy duty ball bearings are fitted and oil lubricated through constant level oiler arrangements. Process pumps for

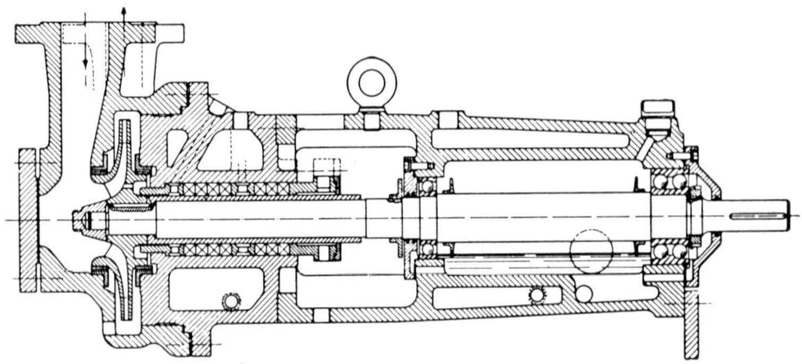

Fig. 6.11 Section through the working parts of a process pump.

larger capacity may be of the double entry type shown in Fig. 6.12. The casing is horizontally split and pedestal supported and can be removed to provide access for maintenance without disturbing the suction and discharge flanges. Oil lubricated bearings are used and the stuffing boxes are usually water cooled.

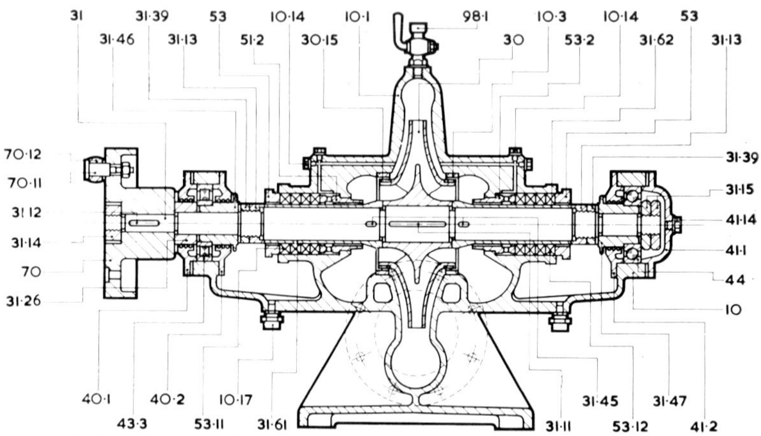

10	Casing, Lower Half	31.61	Shaft Sleeve D.S.
10.1	Casing, Upper Half	31.62	Shaft Sleeve N.D.S.
10.14	Water Sealing Channel Plug	40.1	Bearing End Cover D.S.
10.17	Bracket Drain Plug	40.2	Bearing End Cover D.S.
10.3	Wearing Ring for Casing	41.1	Bearing End Cover N.D.S.
30	Impeller	41.14	Screwed Plug N.D.S.
30.15	Wearing Ring for Impeller 30	41.2	Bearing End Cover N.D.S.
31	Shaft	43.3	Roller Bearing
31.11	Key for Impeller 30	44	Ball Bearing
31.12	Key for coupling 70	51.2	Neck Bush
31.13	Shaft Nut	53	Stuffing Box Gland
31.14	Screwed Collar	53.11	Stuffing Box Packing DS
31.15	Nut for Ball Bearing 44	53.12	Stuffing Box Packing NDS
31.26	Distance Sleeve	53.2	Neck Bush
31.39	Lockout for 31.13	70	Pump Half Coupling
31.45	Keys for 31.61 and 31.62	70.11	Coupling Pin
31.46	Water Thrower, Driving Side	70.12	Coupling Buffer
31.47	Water Thrower, Non-Driving Side	98.1	Air Cock

Fig. 6.12 Split casing pumps (a) Longitudal section of a horizontal double-entry split casing pump.

Fig. 6.12 Split casing pumps (b) External features of a split casing pump.

Two stage pumps for higher heads are available as shown in Fig. 6.13, a single end cover version being shown. This design has the impellers back to back and the overhung arrangement retains the advantages of the single stage pump described above.

Double end cover versions are also available, both having the casings pedestal mounted. Two stage double entry horizontal split casing type pumps can also be used for higher head applications.

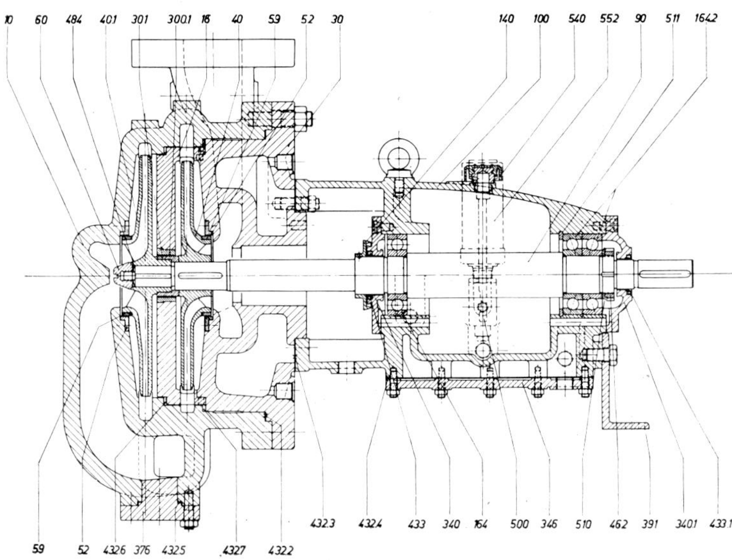

Fig. 6.13 A two-stage process pump.

3. *Multi-stage pumps*

Multi-stage pumps are used for high head applications and are basically of two types, ring section or barrel casing. The ring section pump consists of a number of sections or stages, each of which contains an impeller, diffuser and return chamber, secured between two end covers by means of external bolts. A typical ring section pump is shown in Fig. 6.14.

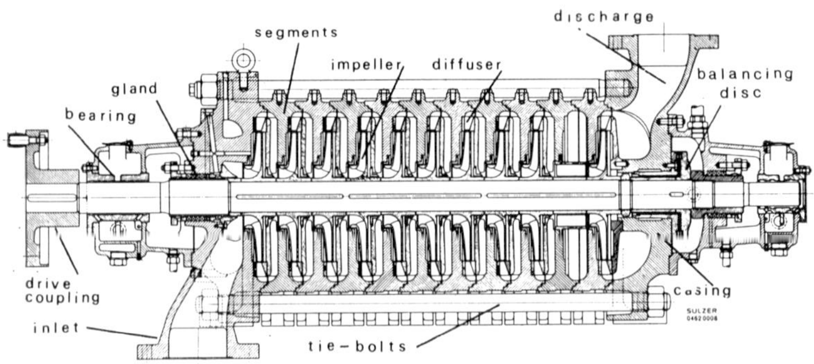

Fig. 6.14 Typical ring section multi-stage pump.

For very high pressures the securing bolts become very large especially on boiler feed pumps where rapid temperature changes have to be withstood without the bolts losing their prestressing. In these cases a barrel casing pump is often used. This is basically a ring section pump installed in a barrel or pressure casing, and only has one external joint which can be sealed by relatively short bolts. Interstage leakage is avoided by a rigid mechanical joint and the presence of water at delivery pressure in the clearance between the end pump segment and the end cover. Temperature variations are sometimes minimised in pumps of this type by the transference of the liquid over the segments of the pump and consequently distortion is reduced to an acceptable limit. A typical barrel casing pump is shown in Fig. 6.15.

10	Volute casing
16	Interstage diaphragm
30	Stuffing box housing
40	Impeller
40.1	Impeller
52	Casing wearing ring
59	Impeller wearing ring
60	Impeller nut
90	Shaft
100	Bearing frame
140	Deflector
164	Deep groove ball bearing
164.2	Angular contact ball bearing
300.1	Bushing
301	Sleeve
340	Bearing end cover
340.1	Bearing end cover
346	Cover
376	Blank flange (drain I)
391	Foot
432.2	Gasket
432.3	Gasket
432.4	Gasket
432.5	Gasket (drain I)
432.6	Gasket
432.7	Gasket
433	Radial sealing ring
433.1	Radial sealing ring
462	Slotted nut
484	Cup spring
500	Oil gauge
510	Locking plate
511	Circlip
540	Oil filling plug
552	Constant level oiler

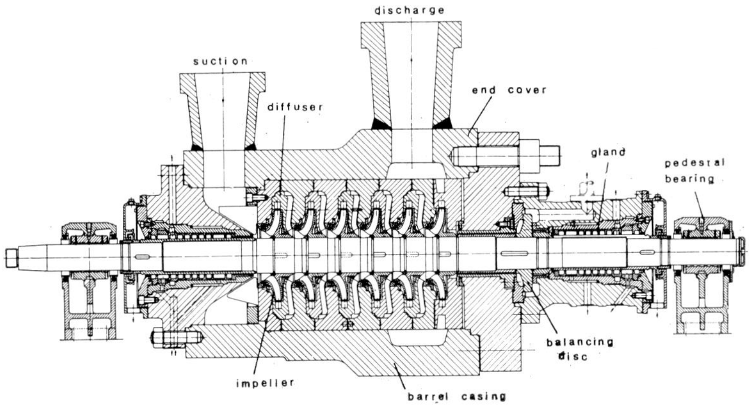

Fig. 6.15 Typical barrel casing multi-stage boiler feed pump.

The majority of multi-stage pumps use unbalanced impellers to avoid complication of assembly and these have, therefore, unbalanced axial forces to be accommodated. This is normally achieved by the use of either multi-purpose bearings at one end of the shaft or a hydraulic balancing disc.

Multi-stage pumps may have either pedestal or bracket type bearing support. The pedestal type bearings shown in Fig. 6.15 are mounted on the base plate separate from the main casing whereas bracket type bearing supports form an integral part of the pump casing (Fig. 6.14). The latter are generally more satisfactory since there is less possibility of misalignment due to casing distortion especially where temperature fluctuations are likely to occur.

Boiler feed pumps appear in forms throughout the whole range of multi-stage pumps from fairly simple models at lower pressures to more sophisticated versions of the high pressure barrel casing type. They are invariably horizontally mounted and may be very high powered. For these units, running at high speeds and with high stage pressures, a suction booster pump is often required in order to provide adequate suction pressure to the first stage impeller. (See Example Two, Section Five.)

Vertically mounted multi-stage pumps may be process pumps or – in the water industry – borehole type of pumps. It should be noted that a borehole needs to be as small as possible due to its cost and therefore even with heads which are not themselves very large a multi-stage design becomes necessary.

Process pumps appear in multi-stage barrel form for mounting in line, where the barrel may take the form of a wet well or be used as a discharge tube as shown in Fig. 6.16. This type of pump may be pit mounted and in such circumstances care is required against leakage where the fluid handled is likely to give

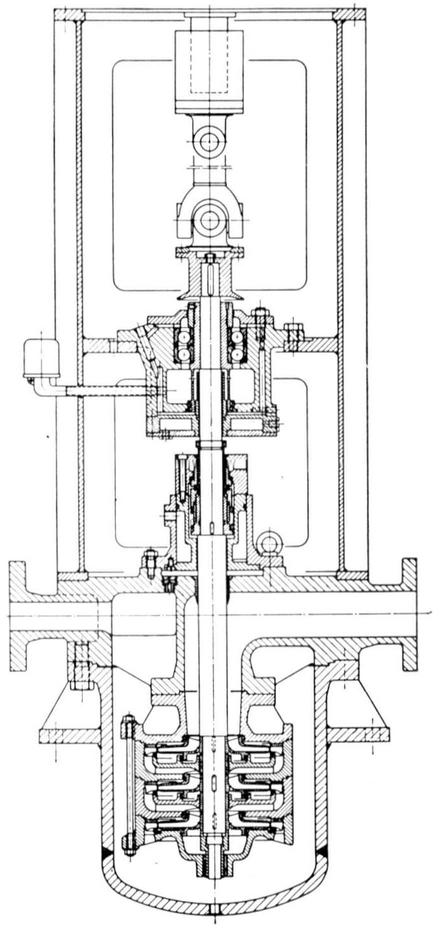

Fig. 6.16 A vertical barrel casing pump.

off heavy vapour. These pumps are compact and share the characteristic of all vertical pumps in that they take little ground area.

Hydraulic balance of the vertical multi-stage pump is sometimes obtained by leading the outlet of the first stage to a central position from which stages work in opposite directions, thus minimising the final out-of-balance thrust to be taken on the main support bearings.

(iv) *Special purpose pumps*

1. *Chemical pumps and acid pumps*

In general these are similar to the simpler process pumps but their design is varied so that casings and impellers may be cast in a variety of materials. Acid pumps often have a ceramic inner casing externally reinforced with cast iron, and a ceramic impeller bonded on to the shaft. In these pumps replacement of the impeller means shaft replacement as well.

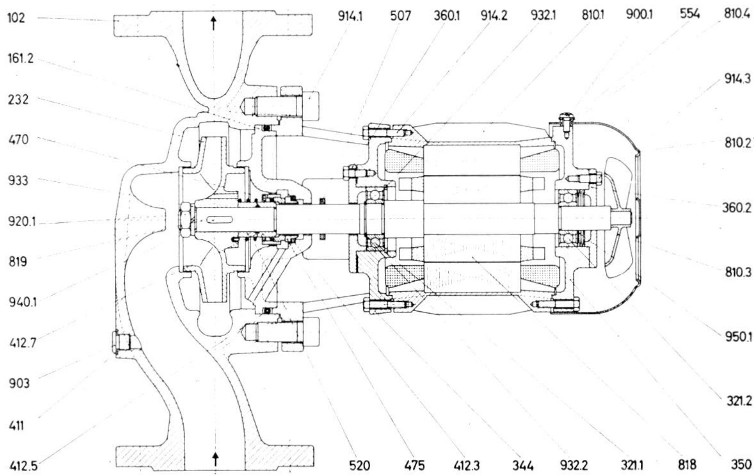

102 Casing
161.2 Cover
232 Impeller
321.1 Ball bearing
321.2 Ball bearing
344 Motor bracket
350 Bearing casing
360.1 Bearing cover
360.2 Bearing cover
411 Joint
412.3 O-ring
412.5 O-ring
412.7 O-ring
470 Parts for mechanical seal
475 Counter ring
507 Throw ring
520 Guiding sleeves
554 Washer
810.1 Stator
810.2 Fan cover
810.3 Fan
810.4 Socket
818 Rotor
819 Shaft
900.1 Fillister head screw
903 Screw plug
914.1 Socket head cap screw
914.2 Socket head cap screw
914.3 Socket head cap screw
920.1 Nut for impeller
932.1 Securing ring
932.2 Securing ring
933 Spring washer
940.1 Key for impeller
950.1 Disc spring

Fig. 6.17 Section through a typical in-line pump.

Leaks are dangerous in such pumps and special precautions to relieve pressure at the glands may be introduced. Pressure relief may be obtained by introducing back-vanes on the impeller or by the use of pressure-operated mechanical seals. Chemical pumps are available in horizontal and in-line vertical form.

2. *Motor mounted pumps*

These pumps are designed for a simple compact assembly and for the lighter duties. The impeller is bolted directly on to the motor shaft extended for this purpose, and the casing on to the motor frame. No bearings other than the motor bearings are needed, but clearly the pumping stresses are transferred to the motor and any breakdown in the one affects the whole unit. Repair is usually on the 'standby replacement' principle often preferred in small, busy plants.

A variation of the above type is shown in Fig. 6.17 where the whole unit is mounted 'in-line' on the pipeline, sometimes

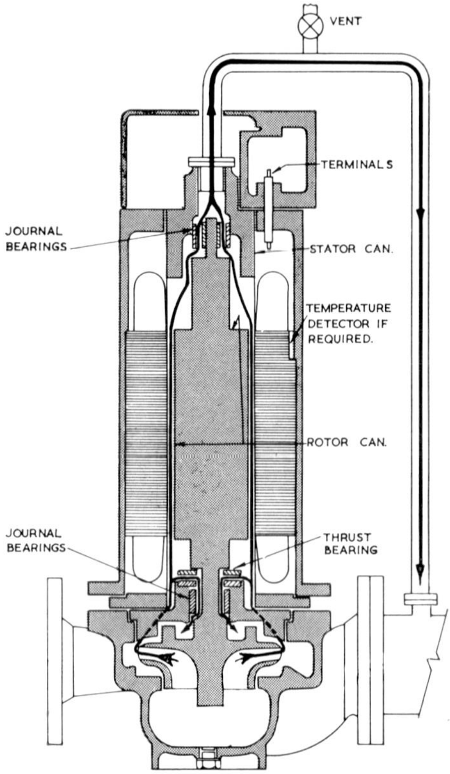

Fig. 6.18 A canned motor pump.

without added support. Central heating pumps are often of this type.

A further variation is the 'canned-motor' type pump shown in Fig. 6.18 where the whole unit is sealed and again mounted in-line with the pipeline. These pumps are leak-proof. The motor is of special design to keep the windings 'canned' and sealed from the fluid. The liquid pumped should have sufficient lubricating properties and should be reasonably clear and non-corrosive since it is used to lubricate and cool the motor and bearings.

3. *Self-priming pumps*

Pumps of this kind are necessary where automatic operation of a pump is required and the pump is mounted above the liquid level. (Foot valves installed in the suction lines of a normal pump so situated are not proof against the pump draining down in due course although they may be useful for short-cycle operation.) This type of pump has an air chamber immediately over the impeller as shown in Fig. 6.19 and the impeller itself is mounted in a trap on the suction branch which ensures that the impeller is always immersed in fluid.

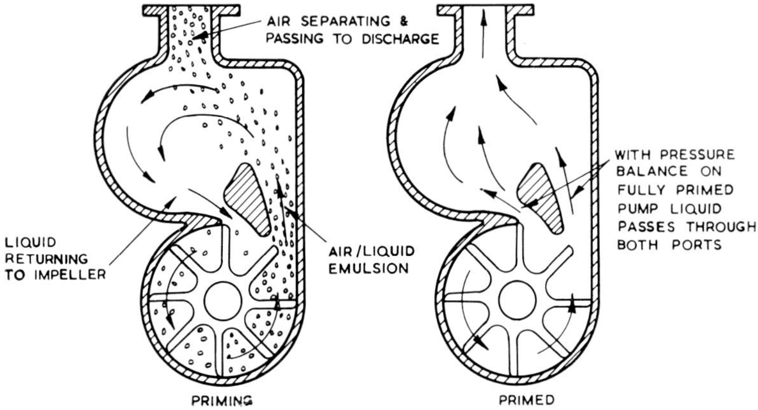

Fig. 6.19 Action of separator on trap and separate self-priming pump.

When the pump is started the air in the suction branch is contained progressively in the residual fluid of the trap and the resulting mixture delivered to the separator where the air escapes into the delivery line and the water falls back again into the impeller until all air is expelled, and the pump is then free to operate normally.

The nature of the design is such that certain precautions are necessary: adequate air venting of the discharge line, careful sealing of the suction line which should also be kept to a minimum volume commensurate with an acceptable priming time. Common suction lines should not normally be used for this reason. Efficiency of this kind of pump is usually low.

4. *Inducer pumps*

These are used where a simple robust impeller is required to handle materials which would clog a normal impeller, and in condition of low head where efficiency is of no consideration. Fig. 6.20 shows such a pump and the impeller action.

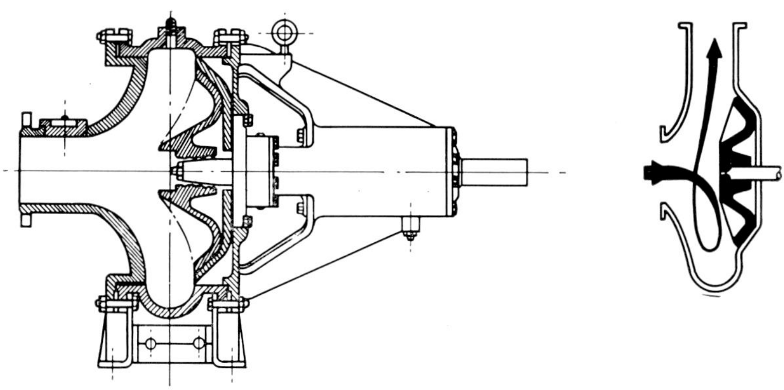

Fig. 6.20 Inducer pump.

5. *Peripheral*

Also known as Turbine, Regenerative Turbine and Vortex. This type has an impeller with vanes around its periphery (Fig. 6.21). They move through a toroidally shaped channel which is restricted at one point between inlet and outlet ports. The liquid is picked up by the vanes when passing the inlet ports and is accelerated outwards and forward into the channel. A vortex is created as the liquid is deflected by the channel wall and returned to the impeller. This cycle is repeated many times, each time adding kinetic energy to the liquid which is partially converted into pressure in the divergent outlet.

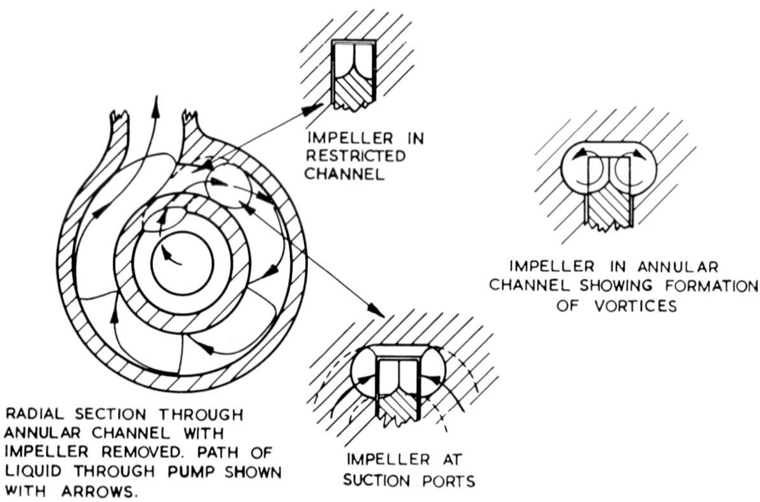

Fig. 6.21 Peripheral pump.

This type of pump is recommended when small capacities (a few cu.m./hr.) against a high head (say 150m.) have to be pumped. A restriction, however, is that the liquid pump must be free from abrasive solids and non-viscous, because of the close running clearances.

6. *Glandless pumps*

These pumps have been developed to meet the need for complete containment of the pumped fluid on the basis of leak hazard or leak cost. E.E.U.A. Handbook No. 26: *Electrically Driven Glandless Pumps* fully describes their type and application.

(b) Operating conditions

(i) *Axial/mixed-flow Pumps*

These pumps are usually installed in conditions where a pump picks up water directly from a wet well, the lowest water-level being above the pump centre-line. Conditions may vary considerably in so far as the head on the system is concerned, from very low head (on, say, drainage pumping systems), to higher heads of up to 12 or 15 m on cooling systems associated with power stations. In these applications, the water level from which the pump is drawing may be subject to tidal variations, and it is often necessary to mount the motor a considerable height above the pump itself, with the pump centre-line in all cases safely below the low water-level.

It is essential in such conditions that the pump be adequately protected by means of screens on the intake, and that there is a sufficiency of cross-sectional area available in the intake at low water, to ensure that flow velocities into the pump are low. Depth of water over the pump in such circumstances should always be sufficient to prevent vortex forming, through which air can enter the pump.

In cooling station practice, these pumps are used in a system containing an elevated condenser and a discharge point not greatly different in level from the intake. In order to take advantage of the lowest pumping head possible in these circumstances, it will be seen from Fig. 6.22 that the hydraulic gradient is situated below the top of the condenser and that the condenser is therefore operating under sub-atmospheric pressure. It is the custom to keep the degree of sub-atmospheric working to no more than about 8 m and where there is a tidal variation at the intake, this 8 m maximum is measured at low tide.

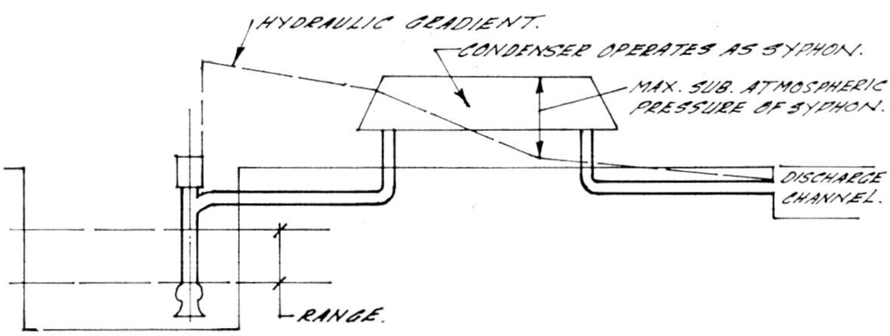

Fig. 6.22 Power station cooling.

With tidal variations, it is clear that the quantity pumped at low tide will be less than that pumped at high tide, and that the system is therefore performing more satisfactorily at the high tide condition than at the low. Such systems are usually designed for a mean tidal condition in so far as maintaining output is concerned (Fig. 6.23).

The mounting of these pumps is of interest in that the discharge elbow and valve should always be above water-level at all states of the tide, unless the wet well can be closed off and dried out. The larger the tidal range, the longer the distance will be

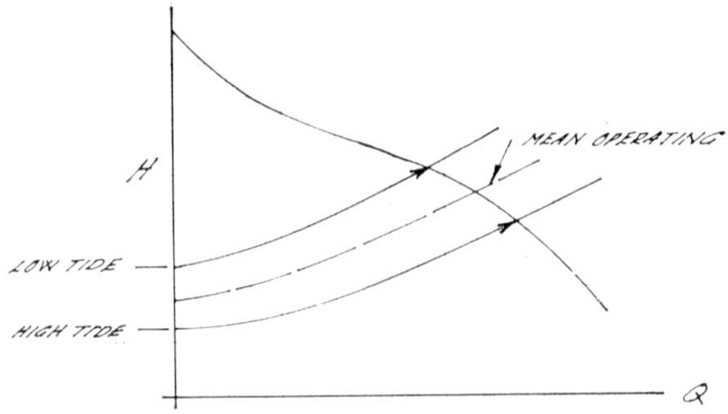

Fig. 6.23 Tidal pumping.

between motor and pump, and head room may, in certain circumstances, become quite large. There is, therefore, a good deal to be said for mounting such pumps in the open, dealing with lifts by means of a mobile crane.

These pumps, when operated in parallel, must be adequately protected by valving. Since the pumps themselves have considerably less resistance in reverse than an equivalent centrifugal pump, it is essential that the valve closes on power failure to a pump, otherwise a considerable part of the delivery from the other pumps will be recirculated through the pump that has failed. It is customary to use butterfly valves inter-connected with the pump electrics in these circumstances.

It is often the case that civil engineering work associated with the installation of wet well axial or mixed-flow pumps shows considerable savings over that associated with centrifugal pumps of the same duty, and this aspect must be considered in making a choice between the two, the overall cost being the deciding factor. It should be noted also that, in the case of mixed-flow pumps, two stages are sometimes used, bringing the head available to a value which would otherwise have required a centrifugal pump to deal with it, and maintaining the advantage of wet well installation.

It could be, particularly in drainage applications, that the total head on the pump is very much affected by tidal or flood variations. In these circumstances, variable speed motors or axial flow, variable-pitch propellers may be economic.

(ii) *Centrifugal pumps*

In industry, centrifugal pumps are so widely used that any generalisation of operating conditions is impossible to make: each installation must be treated on its own merits.

However, it is important that certain parts of the installation be standardised, as much as possible, in any particular factory or group, and it is customary to ensure that flanges are specified to a standard which will enable replacement of pump or valve without adaptation. That is, it is wise to specify the same flange drillings in all pipes, pumps and valves of similar diameter for similar duties.

For large installations, such as mentioned above, centrifugal pumps in dry wells are also often used, and the problems governing their use in such conditions are very similar to those outlined in b(i) above. Centrifugal pumps of this type may be end suction pumps, so large that the entry channels are designed in concrete as draught tubes of the type shown in Fig. 6.24. Pumps of this kind have also been used with concrete volute casings, as economic alternatives to the more usual cast iron. Reliability of operation is essential in all power station applications, and maintenance periods are usually limited to shutdown periods of the station itself. It is not customary to install spare units, due to the size and cost of each individual unit concerned.

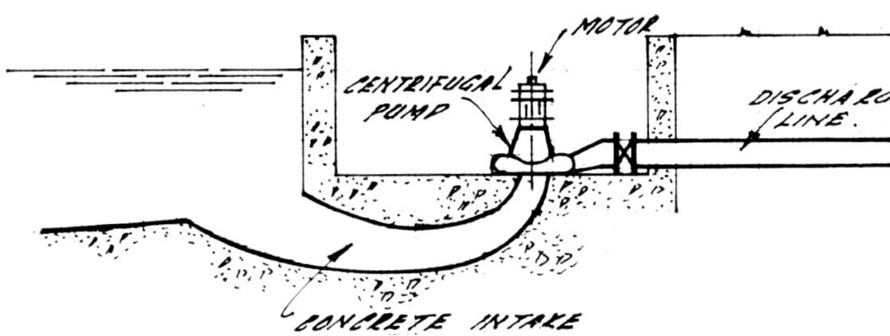

Fig. 6.24 Vertical, volute Centrifugal pump with concrete draught tube.

(iii) *Reliability of operation*

The degree of reliability of operation built into a system will naturally vary with the class and continuity of operation and, except in the case of very expensive pumps, spare capacity is normally considered. Where a spare pump is installed, it is often considered that there should be no rotation of duty of the pumps, since this practice may merely ensure that all pumps become worn out at the same time. A cheap spare pump may therefore be installed in an expensive system as a stand-by, provided it is run only on occasion, to ensure that it is fully operational.

Present tendency is for pump houses not to be 'manned', and consequently some reliable indication of the state of running of pumps is necessary at some central control point. It may be that, in practice, full control of the pumps is needed from the central point. The minimum indications necessary at a central point will relate to the state of running of the pump and the state of its associated discharge valve. In this case, the usual sequence is that, when the pump is switched on, the valve automatically opens as soon as pressure has built to a sufficient head in the pump casing.

Some motors are water-cooled, and also water-cooling of seals on the pump is usually necessary. The supply for such items is often taken from the discharge side of the pump casing and returned to the suction side. In an important installation, it may well be necessary to install instrumentation on these cooling lines, which would show up a fault should this occur.

If, for any reason, therefore, the pump outlet valve fails to operate, some time-mechanism must be installed that cuts the pump out before damage to the unit is caused through lack of pump cooling. A period of 5 min. is usually adopted. Alarms indicating failures in the control room should be such that they are easily associated with the pumps concerned, and such alarms should be self-cancelling when the pump is idle, otherwise operators will become 'conditioned' to alarm lights showing when, in fact, no action is required and, as a result, may not be as alive to alarm states arising which **need** attention.

(iv) *Ancillary supplies to pump units*

It is important that items essential to the proper running of the pumps should not be overlooked in the general design of a system. In particular, water supplies to gland seals should receive careful attention. Generally, such seals require cooling or flushing at reasonable pressures, and ideally, the supply should not be such that it can inadvertently be left off when the pump is running. The best arrangement is to have interconnections between the delivery part of the pump casing and the seal, with a return line from the seal to the suction end of the pump. If the head across the pump is then sufficient to provide the necessary flow across the seal, there is no danger of the seal being damaged through lack of coolant. This arrangement, however, has its limitations. For instance, the liquid being pumped may contain abrasive material that would ruin the seal, or it may contain suspended solids which would block the small passages within the seal itself. In the case of suspended solids, an internal arrangement of this sort may still be possible, if suitable filters are introduced, but they should always be such that they are easily inspected, and, if the pump installation justifies such measures, alarm mechanisms against filter blockage may be installed.

There may also be a separate pumped lubrication or flushing supply arrangement connected to seals, which must not fail while the pump is running.

Such questions as cooling the seals should always be properly discussed at the time a pump is ordered, and suitable cooling arrangements agreed with the pump and seal manufacturers as being adequate.

(c) Fault conditions

Faults on pumps can be classified broadly as follows:

(i) Mechanical failure
(ii) Power failure
(iii) Instrument failure
(iv) Environmental failure

} Each can affect the others.

(i) *Mechanical failure*

Failure can occur in practically any part of a pump, giving total failure of the whole pump unit. Such is the case, for instance, when there is cracking of the pump casing, or a broken shaft, but such failures are rare, and it is more likely that pump failure mechanically would occur gradually, such as due to misalignment and balance, or due to wear and tear, finally arriving at the point when the pump must be taken out of service for maintenance purposes. Causes of total failure mechanically are usually of the kind where a pump has some mechanical obstruction caught within it at speed. Since the impeller passages are normally inadequate to clear an obstruction of any size, there is usually a binding of the impeller on the casing, causing a broken key, broken impeller or broken shaft. If the driver continues to turn, in such circumstances, it will be grossly overloaded and will probably be damaged unless properly safeguarded.

Assuming that the system when in operation is adequately guarded by screens on the suction side, such a failure is only likely to occur when the system is first commissioned, and dirt, tools, etc., have been inadvertently left in the pipe line. It is essential, then, to see that a system is thoroughly cleaned out before commissioning and that pump movement is checked for freedom before the pumps are switched on.

Cavitation dangers have been mentioned already, and although cavitation over a short period is unlikely to be dangerous, continued cavitation can cause serious pitting of the pump impeller, and the condition should be avoided by proper design of the system.

(ii) *Power failure*

Surges in a system due to electrical power failure, which have already been mentioned as possible causes of high pressure breakages in both pump and system, should be adequately guarded against. An electrical failure in any part of a pump's associated electrics could also be serious. For instance, an electrical device indicating flow of water to a seal may fail, and if, in such circumstances, filters are blocked, no alarm will be sounded and seals may be ruined. Some pumps, when dealing with very viscous fluids, may have heaters in the fluid to keep the viscosity at such levels as are easily handled. If such a pump is on stand-by, it is essential that the heater concerned be operative, and a failure here would cause the pump to be greatly overloaded if it is started up. Some materials freeze out in the system in such conditions, and the degree of safeguard concerned will, of course, depend on the nature of the fluid. Again, in the case of stand-by pumps situated in damp conditions, the motors may be fitted with heater coils. It is essential that supplies to the heater coils are maintained, if the pumps are to be kept in order.

(iii) *Instrumentation failure*

Where a pumphouse is unmanned, instrumentation for sensing the operation of the pumps can become complicated, and may cover any (or all) of the following:

- Liquid temperature, flow, and pressure.
- Pressure interlocks for automatic start.
- Low level of liquid in the store and eventual automatic pump tripping.
- Motor temperatures.
- Seal coolant.
- Valves: state of opening and closure.

This instrumentation will have certain interlocks designed to protect both pumps and product, and it must be ensured that an instrument failure does not cause a stoppage of the system which is spurious. For instance, an alarm is perhaps preferable to automatic closure, if there is time for the alarm to be assessed and manual action taken. In this case, it is customary to install a scanner for the instrumentation, which automatically shows an alarm on a scan panel so that action can then be judged by a central operator. An example of 'spurious automatic shut-down' could occur in the case of a low level cut-out operating on an air reaction device. Although the liquid levels may be perfectly adequate, the instrument may cease to operate if it loses its air supply, and will then shut the plant down quite unnecessarily. Useful guidance on control valves is given in E.E.U.A. Handbook No. 32: *The Selection of Automatic Control Valves.*

(iv) *Environmental failure*

This is particularly important where the liquid being pumped is either volatile or corrosive, or where large quantities of water are handled through an area such as a basement that may be occupied. If, in such conditions, there is a pipe failure, pressure on the system immediately drops, and the pump is encouraged to deliver large quantities at lower pressure. It may, therefore, be possible to place a sensing device on the pump delivery which, if the pressure drops below a certain level, automatically switches the pump off. Similarly, in the case of a basement, a switch operated by a rise in water-level can ensure that the pump is switched off as soon as flooding starts. In this case, however, care should be taken that the valves on the system are also automatically closed, especially if there is a possibility of the flood continuing by draining down large storage volumes.

A valve failure on a stand-by pump can cause recirculation through that pump, in which case the pump may be driven in reverse to a runaway speed. Whereas this condition is not usually dangerous, it can cause a loss of product, and is usually indicated by a drop in pressure not explained by any visible causes.

In the case of a pump with a suction lift, a priming device (such as a vacuum line) is usual, and failures of such a line will prohibit the pump being started. If the ejector on the vacuum line is operated by air, some stand-by in the form of a steam line to the ejector is also advisable; or, if a vacuum pump is employed, it may be necessary to put on a duplicate pump. Cases have occurred where, with the vacuum system in good order, pumps have primed, only to lose the priming as soon as the discharge valve is opened. Such cases are usually explained by air being trapped between the discharge valve (which may be a motorised sluice valve) and a check valve further down stream, so that, when the motorised valve is opened, a further quantity of air enters the pump casing, causing the priming failure. Separation of valves in this way should be avoided wherever possible (Fig. 6.25). Where pumps are mounted in a well or a basement, and there is danger of flooding, it is elementary that the motor must be raised to a point above flood level, Where this is not economic, submersible type motors can be used.

Particular care should be taken that a fault cannot develop

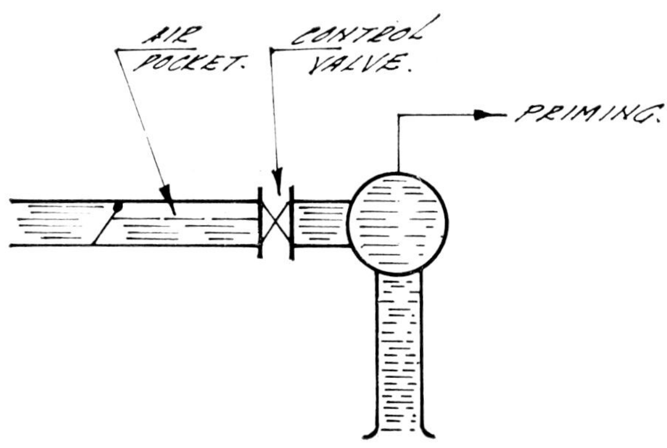

Fig. 6.25 Loss of priming.

which is a 'sequence fault'. In other words, a fault affecting one pump should not by its nature trigger off a similar fault in an adjoining motor (or pump), particularly in the case of valves. Where a valve is common to more than one pump, a good case for its introduction should be necessary and, when it is introduced, it should be of such quality as to make failure very unlikely. Eventually, the security of the design adopted must, of course, be guided by economics, but adequate thought given in the design of the pump, its electrics and its environment can avoid serious losses at a later date.

(d) Materials of construction

A pump for pumping cold, clean water would normally have a cast iron casing and impeller, steel shaft and probaby non-ferrous (i.e. gunmetal) renewable components such as shaft sleeves, sealing rings and gland. If the pumped liquid was other than water (for example, hydrochloric acid), the casing would be of ceramic and the shaft of ceramic coated steel. All other parts in contact with the acid would also be of the same material, with possibly a PTFE seal. This is the criterion when considering the initial cost of the pump: the materials in the pump must be compatible with the fluid being pumped, to obtain a reasonable pump life.

Still more exacting requirements may be met involving high pressures or temperatures (or both), when the materials of construction may be stainless steel or some of the newer exotic materials, such as the titanium-based alloys.

By similar methods the pump may be adapted for the handling of foodstuffs, or pharmaceutical products, where the materials of construction must not, under any circumstances, contaminate the product. A typical example would be a pump used for pumping gin or milk; this would all have to be of stainless steel construction, and designed so that it could easily be sterilised. There are unlimited materials of which pumps may be constructed. The present development of plastics or the older lead-lined pumps present an economic solution for unsuitable corrosive duties, where the base material may be extremely costly.

There may well be situations where the user may find it is more economic to purchase a low-cost pump, where contamination of the fluid is not important and the pump life is short, but may be cheaply replaced with a new unit. A typical list of materials of construction for use with certain common fluids is given in Appendix A of this Handbook. Reference should also be made to the following E.E.U.A. Handbooks: No. 6: *Chemical Resistant Linings for Pipes and Vessels*; No. 20: *Thermoplastics Pipework and Ducting in Unplasticised Polyvinyl Chloride and Polythene*; No. 21: *Chemical Resistant Materials, Part 1 and Part 2.*

(e) Types of seal

In the past, packed seals of rubber, asbestos, hemp, cotton and like materials have served the industry well in sealing rotating shafts. As design progressed, however, higher pressures and temperatures, and the handling of light hydrocarbons and other fluids of serious toxic hazards, indicated the limitations of packed types of seal.

To meet the more stringent requirements, a mechanical type of seal was developed, the principle of which is in all cases the mating of two flat surfaces having a low coefficient of friction, one surface rotating on the shaft and the other stationary, the seal itself being effected by a minute interface film of fluid between the surfaces. Most mechanical seals provide for the sealing surfaces being compressed together by a single spring or bellows (or series of them), or by fluid pressure. Many types are available, depending on the fluid handled, the temperature and pressure conditions, and the shaft rotating speed, and it is a matter of economics whether such a seal is used in preference to a packed gland in the more simple operating conditions. Recent developments in so called 'glandless' designs are based on the hydraulic pressure developed during pumping forming the seal itself. Care should be taken in using this method in view of possible leakage during the run-up of the pump, and unless special precautions are taken during idle periods, the suction fluid level should be below shaft centre-line to avoid leakage at low pressures. More sophisticated pumps of this nature are appearing on the market, and the reader is referred to manufacturers' data for suitable design information. It should also be noted that a comprehensive list of sealing devices appears in E.E.U.A. Handbook 27: *Dynamic Sealing of Rotating and Reciprocating Shafts*.

Seal cooling should always be discussed with the pump manufacturers. Sometimes it is possible to cool the seals by a connection from the pump outlet, through the seal and back to the pump inlet. Such arrangements are compact and need no external water supply. If, however, the fluid being pumped is of poor quality, and contains suspended particles likely to foul seal passageways, either an external clean cooling supply is necessary, or cyclone-type filters must be fitted on the cooling lines from the pump. Mechanical seals are particularly sensitive to fine suspended particles.

(f) Choice of prime mover (pump driver)

(i) *General considerations*

The choice of prime mover must in the end, of course, be determined by considerations of capital and running costs, but in certain cases the lack of availability of existing electric power may mean that other systems of power may be more economic. If an isolated pumping station, for instance, is to be built, and no electric power is available nearby, it may well be that the only possible means of power is a diesel or petrol internal-combustion engine. If, however, the pump is to be powered in an area of ample steam supply this may, as in the case of steam turbine-driven feed pumps, be an economic alternative to electric drive.

It is essential that, if a steam driven prime-mover is employed, then in the case of starting up the plant, the original starting power should be safeguarded, so that, if steam is not available at the start of the operation, some electrical or other drive must be available for the time concerned. The choice will also be affected by the location of the unit, i.e. whether it is installed out in the open or possibly in dirty or corrosive environments. If internal-combustion or gas turbine engines are to be considered, close attention must be given to the storage of fuel and the necessary safety precautions with regard to the installation.

Maintenance may in certain cases be of prime consideration, either due to the degree of maintenance required or due to the remote location of the pumping unit concerned, and if major maintenance is to be avoided, then it is often desirable to keep the running speed of the unit as slow as possible, and the unit overall as simple as possible.

Whichever prime-mover is eventually considered, the load characteristics of the unit must at all times match: that is, both pump and prime-mover must be compatible. In this connection, also, attention must be given to the efficiency of the unit overall. Since the overall efficiency of the complete unit is given by the combined efficiencies of the driver itself, the coupling and the pump, it is important to keep the efficiency of the driver itself as high as possible, since the capitalised value of loss of efficiency may be in some cases equal to the purchase price of the pump.

(ii) *Load characteristics*

Fig. 6.26 shows a typical torque-speed characteristic for an induction motor, on which is superimposed torque-speed curves for the three main types of pump under consideration: centrifugal, axial and mixed flow, being run-up against closed valve.

It will be seen that, in the case of mixed flow and centrifugal pumps, the run-up torques are lower than for the driver, and consequently spare torque exists to cover all starting conditions. However, it will also be seen that, owing to the steeply-rising torque-speed curve of the axial flow pump, a state exists where normally there is insufficient driver torque to start the axial flow pump against the normal closed valve conditions which govern the torque curves concerned. It is therefore essential to take special precautions in the starting of axial flow pumps, usually by starting against partially-open valves or some by-passing arrangements of flow. It will also be noted that the acceleration rate from standstill to normal forward-drive speed will depend on the excess of driver torque over that required by the pump, and may be calculated by consideration of the area between the pump and driver torque curves. Therefore, in the case shown in Fig. 6.26, a faster run-up time will be shown for the centrifugal pump as compared with the mixed flow alternative, and that in turn is likely to be faster than the axial flow pump.

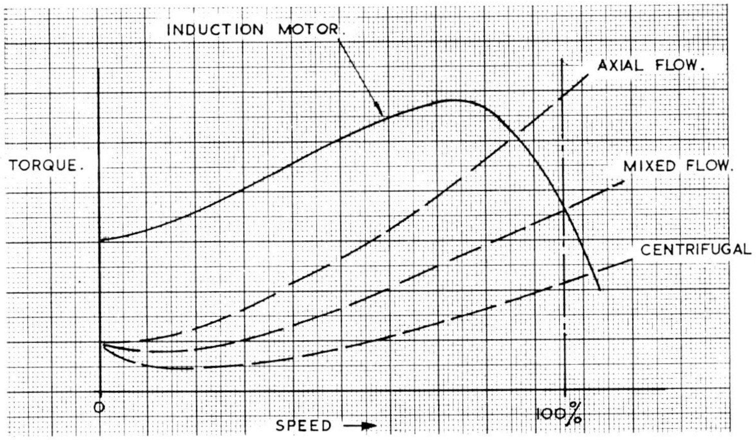

Fig. 6.26 Torque/speed characteristics.

The fast run-up time is very desirable in cases where a pump on stand-by is automatically switched-in to parallel others already running, in order to obtain the minimum disturbance of the system when one of the parallel pumps is lost from any cause.

Fast run-up speeds, however desirable they may be, must be considered against the power requirement of the prime-mover at the start condition, since a high starting torque is usually tied to high currents, with consequent effect on the electrical system.

The power required to drive a pump is equal to the fluid horse power of the pump divided by the pump efficiency plus an allowance for power-loss in couplings and gearbox. In deciding the probable power requirement of the prime-mover itself, a further allowance must be made for a loss of efficiency in the prime-mover, and any spare capacity that may be advisable on account of wear and tear during the life of the plant, which may eventually cause a reduced pump efficiency. The value of this margin depends on the size of pump and the conditions of installation, but for most medium and large size pumps a margin of 10% is usually sufficient. For fractional horse power pumps, however, the margin allowed is usually well in excess of this figure. Manufacturing considerations may also be overriding in the degree of spare capacity obtained, since in the case of electric motor drive, for instance, it is often necessary to use the nearest frame size available, and it may be that the power margin as a result will exceed the 10% mentioned.

In some processes, the materials being pumped are such that in a static condition they may be very viscous. Any considerations, coupled with a similar condition resulting from seal 'stiction', should be considered when determining starting power margins. Since the ouput of a pump depends on its speed, and it is generally desirable that a given change of torque of the driver should have minimum effect on the speed (in other words, that the speed of the driver should be almost constant over its load range), it is customary to choose a driver unit torque curve which is fairly steep in the operating area, or one which is well-defined at its crossing point with the pump 'torque-speed' curve. This calls for a driver having a ratio of full-load speed to no-load speed approaching unity.

(iii) *Types of prime-mover*

1. *Electric motors*

Are very convenient power units for pump driving, requiring very little attention, and being suitable for automatic control and variable speed when required. A motor is so designed that, when correctly installed, used and maintained, the final temperature of the machine is limited to a value which will give the insulation a thermal life of many years. Correct installation and use involves selecting a motor of adequate starting torque and horsepower, and of the right voltage and frequency. The motor casing should be of a type that will provide adequate protection, having regard to the conditions under which it is to be used. The environment in which the motor is to be used must be considered since, if it is likely to be corrosive, hazardous or outdoor, the motor must comply with climatic conditions and statutory regulations.

Most pumps require a motor that is continuously rated and can be run continuously on full-load without overheating. However, if a pump is merely required to run for a limited period at a time (say, up to half-an-hour or so) and will then shut down for a sufficiently long time to enable it to cool down again, it may be practicable to use a smaller motor having a short-time rating.

There is provision for such motors in B.S. 2613: *The electrical performance of rotating electrical machinery,* but there is no really comparable relaxation in British Standard ratings for control equipment.

Because of all the electrical factors to be taken into account in deciding whether or not the use of a short-time rated motor is practicable, these considerations must be judged by the competent specialist. In any case, there would be little saving in the case of small pumps and it is difficult to envisage the installation of a large pump that was only to be run for the limited periods indicated.

A further point to be considered is the number of starts per hour required, since this may cause overheating and/or cause disturbance to public supplies in the case of very large motors operating on 3·3 kV and above.

In deciding on the supply voltage for A.C. motors sizes below 150 kW are generally low voltage, e.g. 415 volt. Normally, 3·3 kV is not used below 150 kW and 6·6 kV not below 300 kW Motors of 750 kW or more are generally 6.6 kV or 11 kV.

In general, high powered motors will be of a smaller size and cost if of a high speed design, and is it not unusual, therefore, for high horsepower units to be offered in conjunction with a gear reducing the final speed to that required by the pump. The choice of motor speed, and whether or not to employ a gear unit, will be influenced by capital cost and overall running efficiency, taking into account the likely pattern of loading throughout the plant life.

It cannot be too strongly emphasised that, in all aspects of electrical work associated with pump operation, the advice of the appropriate electrical specialist be sought, since bad motor choice can affect backing circuit costs very considerably. As an indication, electrical supply and control costs can (and often do) far exceed the costs of the pump/motor unit.

(I) *A.C. motors*

A.C. motors are available in various types, the type choice being generally one that is governed by some special requirement of the case in question. Generally, if two otherwise similar A.C. power units are offered, one having a higher power-factor than the other, the use of the higher power-factor motor will result in lower running costs in terms of electricity charges, and the savings concerned may be quite substantial.

If a variable-speed electric motor drive is required, reference should be made to E.E.U.A. Handbook No. 29: *Electrical Variable Speed Drives*.

The most popular type of A.C. motor is the squirrel-cage induction motor, since it is reliable, robust, efficient and reasonably inexpensive. Its main disadvantage is the fairly high starting current required.

The slip-ring induction motor can be used as a variable-speed drive if required, although it has a low efficiency when so used. Its main advantage is that it has a lower starting current than a squirrel-cage motor, but generally it is more expensive than the squirrel-cage type and has the added disadvantage from a maintenance point of view of sliprings and associated brush wear, etc.

The main advantage of the A.C. commutator-type of motor is that it can be used for variable speed with a reasonably high efficiency. However, it is costly and subject to commutator and brush wear.

Synchronous A.C. motors are sometimes used, particularly where power-factor correction is required. They have the disadvantage of requiring a means of excitation, and are generally uneconomic for sizes below 750 kW.

The speed of 3-phase A.C. motors is given by $60\,f/p$ r.p.m., where f is the supply frequency and p is the number of pairs of poles. Since p can only be a whole number, synchronous motors on a 50 c.p.s. supply are designed for 375, 428, 500, 600, 750, 1000, 1500 and 3000 r.p.m.

In the case of induction motors, they run at a speed slightly below synchronous speed dictated by load conditions and motor design. Corresponding speeds of induction motors may therefore be about 350, 400, 465, 558, 697, 930, 1470 and 2790 r.p.m., values that may vary considerably from manufacturer to manufacturer.

(II) *D.C. motors*

D.C. motors are sometimes specified either due to conditions where speed variation is of importance, or where the security of the system is such that the driving unit is run from a battery power supply. Under failure conditions of mains power, the unit is then automatically transferred to the D.C. supply source. Equally, their use is generally limited, because D.C. supplies are not normally available.

2. *Steam turbines*

If a suitable steam supply is available, the steam turbine can prove an economical method of driving a pump, and offers an easy method of speed and load control. It is customary to use the exhaust steam from the driver for heating or some other purpose, and in this manner it is possible to utilise above 80% of the heat in the fuel, instead of 30% or so when the latent heat is wasted.

3. *Compression-ignition engines (diesel and dual-fuel)*

Dual-fuel engines are basically diesel engines adapted to burn mainly gas with a pilot injection of oil for ignition. They can be run on 100% oil. They are more costly than diesel engines of similar and equal output, because of the additional equipment required, and because the available bmep is reduced when burning gas (so the total cylinder volume has to be increased to compensate).

Diesel and dual-fuel engines can be economically employed over a very wide power range, but, to justify use in any particular situation, advantages must be found to outweigh the following disadvantages relative to electric drives.

(I) *Disadvantages*

- They are very expensive to purchase, install and maintain.
- Considerable problems may arise from noise, vibration, physical size and disposal of exhaust.

(II) *Advantages that may be relevant include:*

- Independent of electrical supplies.
- Fuel cost per hp hour is usually much less than for purchased electrical energy, especially with large engines designed to use heavy oil fuel, or dual-fuel engines using by-product gaseous fuel.
- A use for heat from the exhaust and cooling system may be found. The heat recoverable economically may be roughly equivalent to the mechanical power developed.
- Their ability to vary speed may be useful.

  It is not possible to define sharply the field of application of compression-ignition engines, but it is likely to be in one of the following categories:
- Where the site is remote from electrical supplies.
- In small- or medium-sized mobile pumping units.
- When essential services have to operate after failure of electrical supplies.
- When the annual operating time is high, and sufficient use can be made of the advantages listed above, they may be more economical than a drive by purchased electricity, even when adequate electrical supply systems are available.

4. *Petrol engines*

   The use of petrol engines can be justified only in small, mobile pumping units, and even in these applications diesel engines are frequently a better choice, because of the properties and cost of petrol.

5. *Gas turbines*

   Are available in small pumping units burning light diesel fuel. The units are compact, with a high power-to-weight ratio (but low thermal efficiency) and are attractive when mobility is important and annual running time is low. Because of the low thermal efficiency, simple gas turbine cycles cannot be an economic choice for high annual running time, unless a cheap fuel is available or there is a large process heat requirement to utilise exhaust heat. Residual oil fuels are unsuitable. In the United Kingdom, the fuel most likely to make such machines economic is by-product gas (such as sewage gas), but even where this is available, other plant may be preferable.

(g) Selection of pump for specific duties

When the general duties of the pump concerned are known, the process of selecting suitable pumps may be facilitated by reference to the pump selection charts that every manufacturer provides to cover the full range of a particular type of pump.

The general type of pump for various duties is shown in Fig. 6.27. Within this broad classification, a chart covering a range of centrifugal pumps is given in Fig. 6.28. It shows which pump outline dimensions correspond to a pump duty point and which duty points may be covered by any pump of given outline dimensions. It is presented for a speed of 2900 r.p.m. and single stage pumps, and is reproduced from B.S. 4082, Part 1.

The chart is constructed from a minimum capacity of 3 m^3/hr., progressing in geometrical ratio of 2 up to 384 m^3/hr. It covers head variations from 10 to 160 m. in geometrical ratio of $4^{\frac{1}{3}}$.

These ratios are derived from the need to cover the entire range by a minimum number of specific speeds and this is fulfilled from the capacity ratio 2 by

$$\frac{\sqrt{Q}}{H^{\frac{3}{4}}} = \frac{\sqrt{2Q}}{(xH)^{\frac{3}{4}}}$$

from which

$$x = 4^{\frac{1}{3}}$$

The corresponding specific speeds progress in a ratio of $\sqrt{2}$ from 15·8 to 126·4 (metric). This is since the specific speed with a capacity variation of 2 is in the proportion

$$\frac{n_{s1}}{n_{s2}} = \frac{\sqrt{2Q}}{\sqrt{Q}} = \sqrt{2}$$

Each block such as the area of 30 F on the chart shows a preferred area of operation – it does not cover the full range of any impeller in that particular casing. It does, however, show the area in which that pump is likely to maintain reasonable efficiency. The zone lines are based on stable pump characteristics giving about 15% rise of head to a 50% capacity change.

Charts such as these are built up in different ways by different manufacturers, and the chart shown is merely one example of a rational approach to the problem.

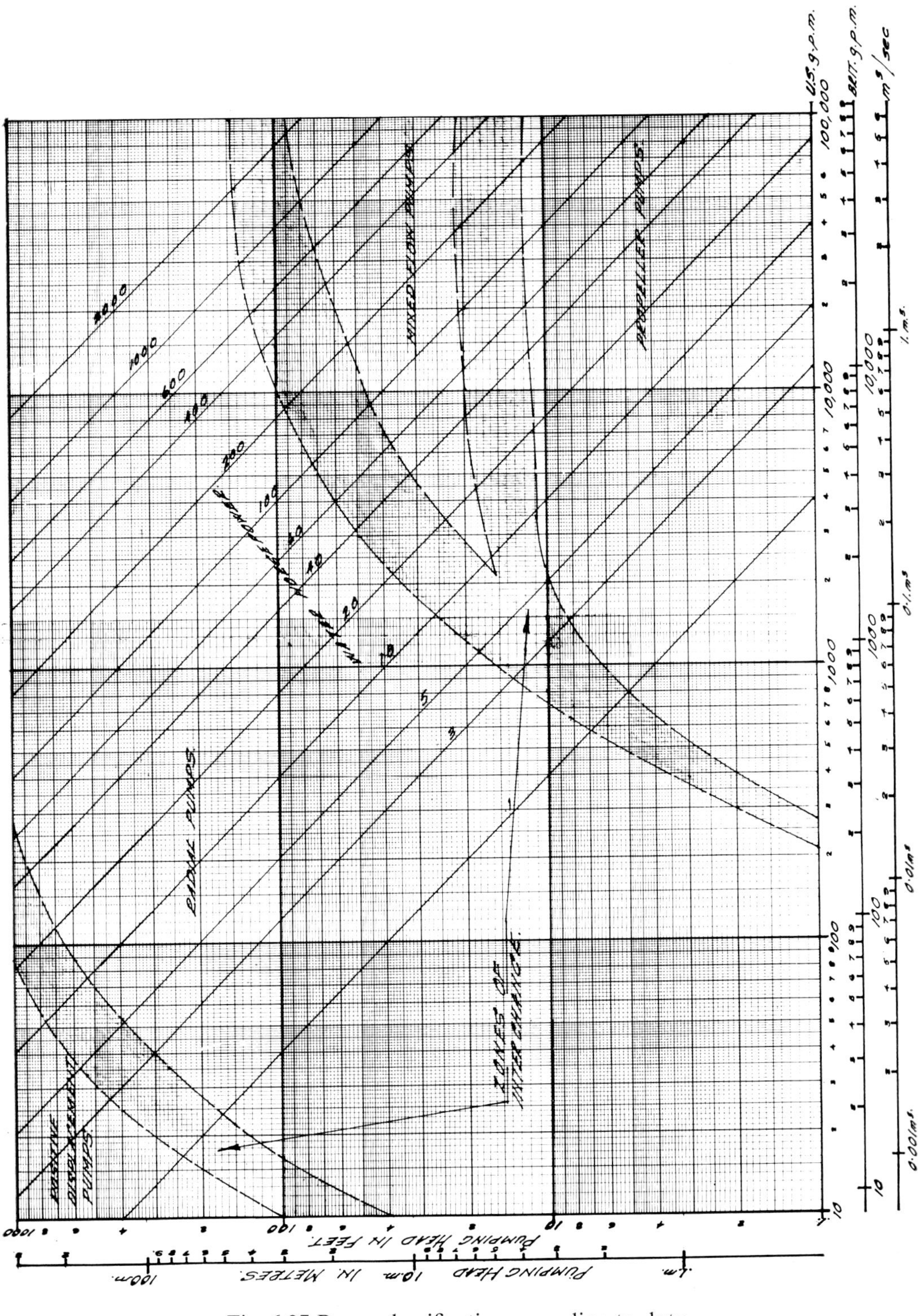

Fig. 6.27 Pump classification according to duty.

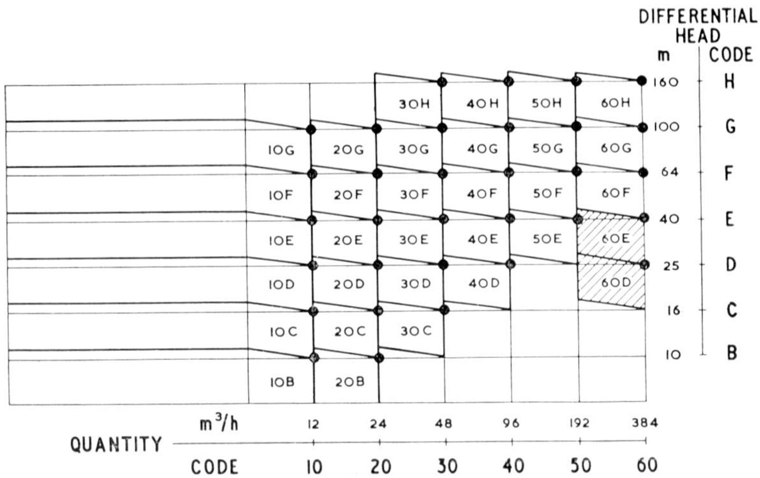

NOTE: This chart is based on a rotational speed of 2900 r.p.m. except for pumps designated 60D and 60E, which are based on a rotational speed of 1450 r.p.m.

Fig. 6.28 Basic selection chart for pumps.

(h) The selection process

It will be appreciated that the proper selection and procurement of a thoroughly suitable pump will involve a knowledge of all previous sections of this handbook, and it is intended now to review the stages through which the selection process is reached, as follows:

(i) A thorough understanding of the conditions of operation that the pump (singly, in parallel or in series) has to meet. This means a proper investigation of head quantity and suction conditions, and the nature of the fluid pumped.

(ii) Having established duty point, the most likely running speed and the specific speed, the type of pump needed will be apparent.

(iii) The operating conditions must be stipulated: continuous or intermittent, atmospheric state, special materials, self priming, etc., and whether it has to match existing pipework or pumps.

(iv) The above data is incorporated in a broad specification which may, if necessary, be the means of obtaining preliminary quotations and further details of proposed pumps from manufacturers. At this stage, also, by reference to selection charts, likely pump sizes and ranges will be apparent.

(v) Finalise a tender specification and invite tenders, or choose from standard stocks. Add requirements for delivery, testing and essential specification references to drivers or prime-movers, such as available power supply or any limitations on such a supply.

(vi) When competitive tenders are received, ensure that all the information is presented on an even basis and that the information is directly comparable from one tender to another. Obtain any further information necessary.

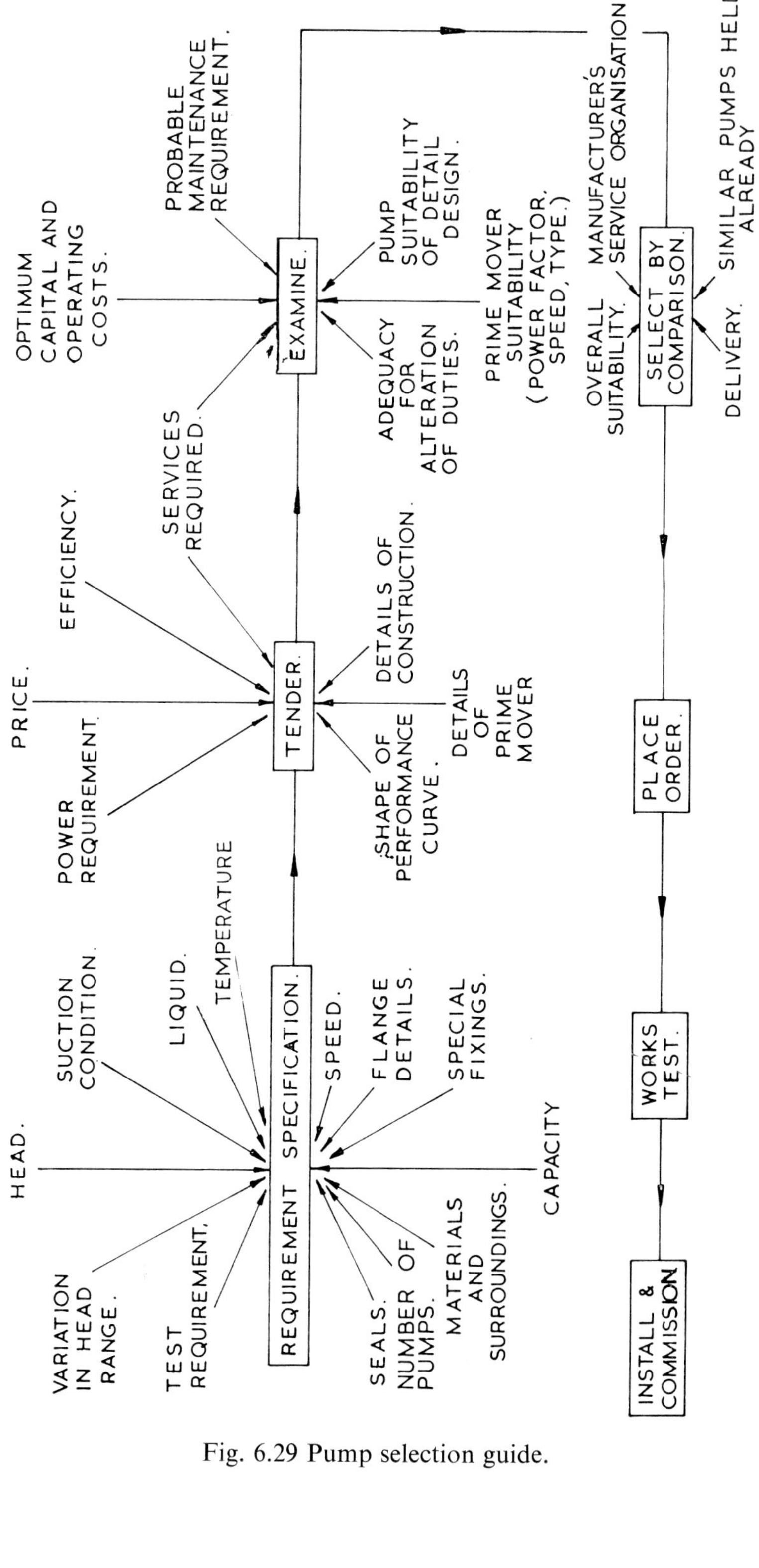

Fig. 6.29 Pump selection guide.

(vii) Examine and compare all tenders thoroughly on an equitable basis, and derive the most economic tender that is technically suitable. Examine also, at this stage, whether any minor amendments may bring about price reductions and, on the other hand, whether casing sizes as tendered are really adequate for future extensions of plant.

(viii) **From this examination, select the best pump.** Even the most economical, on the basis shown in (vii) above, may have disadvantages if previous pumps of the same kind have been bought from a different manufacturer, for different spares may therefore have to be stocked. Consider, also, the probable maintenance requirement, and whether the manufacturer has a suitable organisation, should need arise. Also, whether delivery is likely to be critical, and whether a delivery promised, which is only marginally acceptable, is likely to cause additional costs should delays occur.

(ix) **Only when all these aspects have been considered should an order be placed.**

The flow of information appears on the **Pump Selection Guide** Fig. 6.29.

SECTION SEVEN

INSTALLATIONS

(a) Plinths, baseplates and couplings

(i) *Plinths*

Wherever practicable, the pumping unit should be securely bolted to a plinth, which should be high enough to permit easy access to drain plugs for free draining of the contents of the pump and pipework. The plinth should be designed to allow good access for checking the routine maintenance. It may even be large enough to avoid any overhang of the pumping unit, and may even be extended to minimise damage to the unit from trucking or other traffic.

It is essential to furnish baseplate dimensions to the plinth designer at an early stage, particularly when the plinth has to be tilted or otherwise protected from corrosive fluids being pumped. A bolting template should be used when necessary, and it is essential that the protective tiles or coating should fit closely around the holding-down bolts.

When the pump and motor are dowelled in-line to a common baseplate by the manufacturer, this alignment should be maintained (by the use of tapered wedges beneath the baseplate), when mounting on the plinth. When the correct alignment has been obtained, the wedges must be replaced by solid steel packing plates of identical thickness, prior to the bolting down and grouting. If the motor is not dowelled, axial alignment can be adjusted by moving or shimming the motor, and angular alignment with wedges. The motor should be dowelled after the grout has set. When a pumping unit is mounted on structural steelwork, correct alignment should be obtained by shimming the baseplate feet as necessary, and then be rechecked after several days' running.

Pumps should not be mounted on open steel flooring, due to leakage dripping below, but if this is unavoidable, drip trays should be considered.

(ii) *Baseplates*

Baseplates are generally made of cast iron or fabricated steel, cast iron having better corrosion resistance and usually being cheaper. It is normal practice for the manufacturers to supply the pump and motor mounted, aligned and dowelled on the baseplate, but where the customer is supplying the motor, the necessary dimensioned drawings must be provided.

There are many practices in common use where the purchaser will order a bare shaft pump, pump on baseplate, pump on baseplate with coupling, and it is the customer's responsibility to specify, in detail, what is required in this respect, to include painting or any other anti-corrosive treatments.

(iii) *Couplings*

Types and details of couplings that may be used are to be found in Appendix B.

(b) Pipeline connections

(i) *Suction pipework*

Suction pipework should be as short and as straight as possible, consistent with plant layout. Where the pump is below the suction liquid level, the pipework should fall continuously to the pump; where it is above, the piping should rise as smoothly as possible to minimise cavitation and air removal. Where the pump suction flange is a different size from the pipework, an eccentric reducer should be used, oriented as necessary for air removal.

(ii) *Filters*

Temporary filters should be fitted on the suction side during commissioning, and permanent ones fitted on all ceramic and tanker off-loading pumps, where there is a possible hazard due to ingress of tramp materials. This applies also to suction tanks, which may be open-topped.

(iii) *General*

The pipework, valves and other ancillaries should be designed, fabricated, fitted and supported so that the minimum forces are transmitted to the pump, a typical arrangement being shown in Fig. 7.1. Some loading of the pump casing is inevitable and Fig. 7.2 shows typical allowances made in cast steel case designs for this purpose. Reference should also be made to E.E.U.A. Handbook No. 18: *Supporting of Mild Steel and Cast Iron Pipes,* and E.E.U.A. Handbook No. 20: *Thermoplastics Pipework and Ducting in Unplasticised Polyvinyl Chloride and Polythene.*

(c) Safeguards

(i) *Mechanical*

It is incumbent on the designer and user that all rotating parts should be completely guarded in every respect. (See Factories Act 1961, Sect. i.)

Some companies pay more attention than others to this aspect of safety, but the old maxim, 'It is better to be safe than sorry', should be considered.

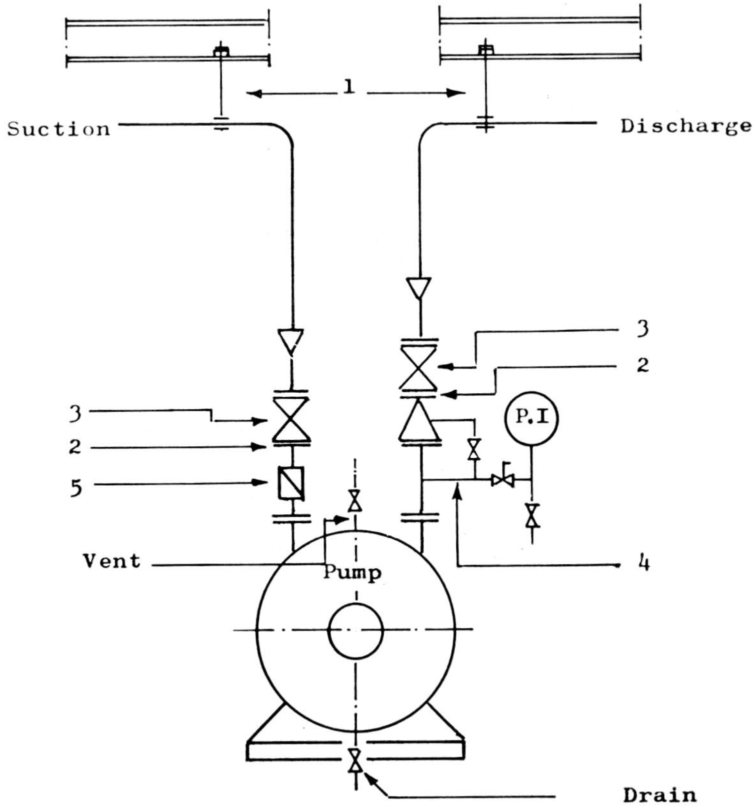

Fig. 7.1 Relationship of adjoining pipework and fittings to the pump

Consideration should be given to the fitting of surge pots on long delivery lines, e.g. drum or tanker filling stations.

Adequate protection against freezing and radioactivity are of vital importance.

(ii) *Electrical*

Adequate comment on electrical safety cannot possibly be given in this Handbook, and is not intended to be covered here. The advice of the electrical specialist should always be sought.

(d) Typical layouts, lifting arrangements and headroom

Where a pumping unit (or any part) weighs more than 25 kg and it is necessary to remove it for maintenance, a lifting beam or other facilities should be made available.

Headroom should also be available for the removal of part or all the unit for maintenance.

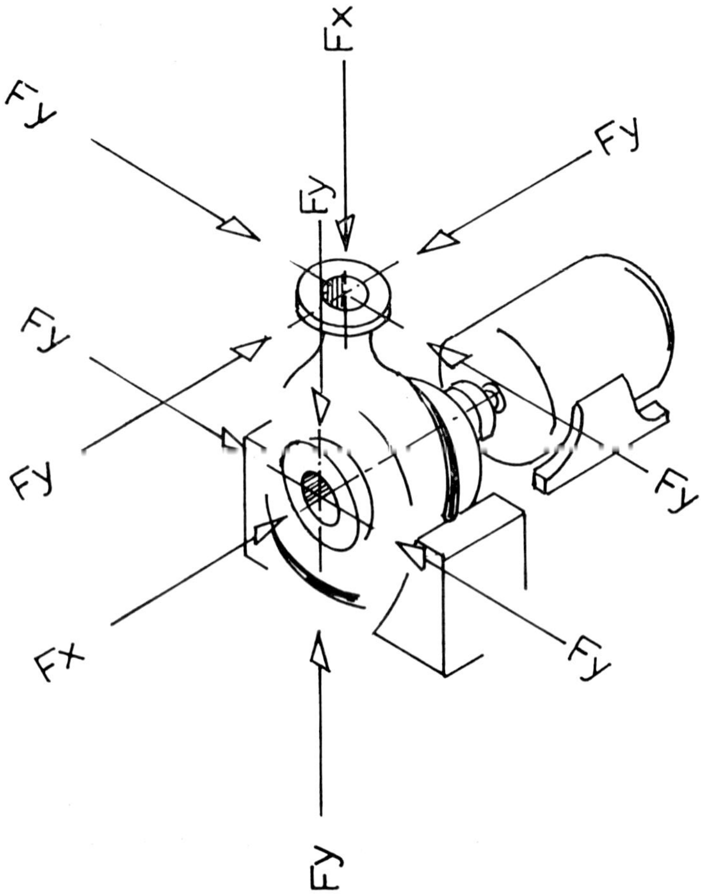

NOTE: The allowable bending stress on pump nozzles is 25% of the maximum allowable piping stress.

F = Reaction forces (kg.). Fx = Allowable = 150D
D = Nominal sizes (in) Fy = Allowable = 50D

Fig. 7.2 Allowable stresses for cast steel pump.

(i) *Valves*

Where a non self-priming pump is used, or where a pump operates under frequent stop-start conditions with a suction lift, a foot valve should be fitted to the inlet end of the suction piping. The valve should preferably have the same dimensions as the pipe suction flange, or have a spool-piece fitted between the pump section flange and thevalve.

Consideration should be given to fitting an isolating valve of the straight-through type to the delivery branch of the pump. Only in the absence of any control valve on the delivery side should the pump be controlled with this valve.

If an automatic control valve is fitted, the engineer responsible for selecting the valve should be consulted, before selecting the pump and finalising the pipe design. For manual control valves, a suitable type should be used, preferably diaphragm or needle.

A syphon breaker or a non-return valve should be considered, where there is a possibility of syphoning back through the delivery pipeline when the pump is stopped.

When two or more pumps are normally running in parallel, or where there is high pressure discharge, consideration should be given to fitting non-return valves on each pump delivery.

Where non-return valves are fitted, suitable means should be provided for draining pipework on the discharge side of the valve. A by-pass arrangement may be required for draining purposes.

All valves, particularly delivery valves where several pumps discharge into a manifold, should be positioned so that they are readily accessible. Alternatively, extended spindles, chain drives, etc. should be used.

Distance pieces should be fitted between the suction and delivery valves and the pump. These can readily be withdrawn, leaving ample space to remove the pump without disturbing the main pipework. In corrosive conditions, the distance-pieces can usefully be fitted with drain valves, gauge and seal circulating connections.

The diameter of the pump suction and delivery branches should not be assumed as the correct pipe diameter.

(ii) *Pumping radioactive liquids*

When radioactive liquids are pumped, the degree of radioactivity of the material concerned will determine the precautions to be taken. Certain elementary precautions are observed, whatever the degree of activity, as follows:

1. The pumps should preferably be housed on a different (that is, lower) floor than the motors and control gear, so that any spillages etc., may be contained in an area that can be separately decontaminated.
2. The pumps should preferably be located above the tank, so that the liquid contained cannot drain down to an open area, as the result of an accident to the pumps.
3. The pump should therefore be self-priming or vacuum-primed, and should also be of good quality to minimise maintenance which, due to decontamination, can be expensive.
4. Pipelines should always be self-draining on completion of pumping, to avoid continuous radiation conditions in the area of the pipes. Such pipelines should also preferably be double-contained, especially where prone to accidental damage. Joints in pipelines and pipelines to pump should be such that the very minimum of liquor is retained in the joint when the pipe is drained down. Ideally, a flush butt-joint on the inside of the pipe is desirable.
5. Where highly-active liquor or sludge is being handled, both pipeline

and pumps need to be heavily protected and, once used, it may not be possible to maintain any part of the system. It is therefore desirable to use the least number of mechanical parts, and such systems are usually free of valves, with pumps of stainless steel or some other robust metal subject to the minimum of corrosive action. Both pumps and motors should be well over-sized for reliability. The installation should also be such that a spare pump can be installed, or has already been installed initially. Faulty pumps cannot normally be maintained, although it may be possible under controlled working conditions for pumps to be replaced.

(iii) *Pumping hydrocarbons and solvents*

Apart from the obvious precaution against leakage which could cause serious fire hazards in themselves, it should be noted that high velocities of flow can be dangerous in terms of static electricity generation, unless the pipework and pumps are well bonded and earthed. Reference is made to E.E.U.A. Handbook No. 26 *Electrically Driven Glandless Pumps* which gives further information on handling low flash-point liquids.

(e) Noise and vibration

Both noise and vibration tend to become standard diseases of machinery. Noise, although more harmful to the surroundings than to the plant, could, in certain circumstances, indicate faulty conditions of operation, whereas vibration is inevitably deleterious to the machine, causing or accentuating misalignment and rapid deterioration of performance.

(i) *Noise*

The energy content of the noise emitted by a pumping unit is a very small part of the total power input. This can be readily appreciated, for instance, when compared with the capacity of a sound amplifier of, say, 50 watts, which is sufficient to be deafening in a moderately-sized hall. Noise is thus negligible in considering the efficiency of the p[ant and so far, therefore, has been given scant attention by pump designers, particularly since its attenuation at source is a relatively costly business, as well as being extremely difficult to achieve. Generally, the noise power level varies directly with the plant horsepower, and within this broad statement come shades of annoyance-levels depending on the spectrum of the noise concerned, from deep rumbles to high-pitched whines.

Noises which are not necessarily deleterious are those occurring with the plant in normal proper alignment and working within its usual capacity. These form the basic level of noise associated with the plant.

Any change in such noise levels normally indicates that some part of the machinery is wearing excessively, or is subject to different (and probably incorrect) conditions of operation. To this extent the noise indication should always be investigated as a maintenance or operating problem. A classical example is the pronounced noise resulting from pump cavitation.

The basic noise level in itself may, however, lead the operator into difficulties. Domestically, if there is labour in the plant room, there could be, in extreme cases, a health hazard. Short of this, the ability to concentrate and carry out duties properly decreases rapidly in uncongenial surroundings. If the plant is to be situated outside, the problem may cease to be domestic and fall into the category of 'annoying one's neighbour', and the operator may then become liable in law for causing a nuisance. The legal remedy for this is for an injunction to be granted to the neighbour against continued operation of the plant. Such a case could cause considerable financial embarrassment both legally and in loss of production.

Local By-Laws governing noise nuisance are generally inadequate, but recent investigation into the whole problem of noise has resulted in more detailed recommendations which, although not at present carrying statutory obligations, may in future be so empowered. This investigation forms the text of the Wilson Report, to which the reader is referred for detailed information.

Indications are that the levels to which any noise must be attenuated are quite low, and if plant is to operate in the open, consideration must therefore be given to:

- The basic noise level.
- Attenuation due to distance away from point at which nuisance is likely to arise.
- Geometric arrangement of plant, so that directivity of noise can be away from the sensitive area.
- Shielding of the noise at source.

A good deal can be achieved by sensible plant arrangement and geometry, and this should always be investigated when there is danger of a noise problem being created. Attention to this when plant is in the design stage is usually a great deal less costly than its correction later.

In the case of small units running in office or domestic installations, quiet running is very necessary, and small central heating pumps of the in-line type, when suitably supported, are very satisfactory in this respect. Larger units, however, nearly always require additional sound insulation, and care must be taken that sound baffles do not adversely affect the cooling air flow to the motors.

Reference should be made to E.E.U.A. Handbook No. 25: *Measurement and Control of Noise*.

(ii) *Vibration*

Generally, vibration falls into three categories:

- Vibration due to misalignment or distortion of rotation parts.
- Vibration caused by insecure mountings to the base plates, or inadequate foundations.

- Vibration as a result of sympathetic periodicity of a supporting platform, or insecure fixing of supports.

In the first category (also contributing to additional noise) falls worn or slack bearings; a badly-aligned shaft; inadequate couplings between pump and motor; a badly-balanced rotor (particularly at high speed); and intermittent contact between moving and stationary parts, such as would be caused by eccentric castings or bores. A gland or seal will cause vibration and noise, if the packing is too hard or is lacking in lubrication, and inevitably leads to further damage. The alignment of motor pump and coupling is most important, and it is essential that, in spite of the presence of a flexible coupling, the general linearity should be as good as possible. Guards, where installed, should always be firmly fixed and sufficiently clear of the couplings as to avoid undue wind noise or physical contact.

It should be noted that it is useless to pay particular attention to all the above matters when installing a pump, only to pull the whole assembly out of alignment when the associated pipework is connected up. Closing pipework should always be eased up to the pump, and not vice versa. A running check after pipe connections is essential.

Vibration due to insecure mountings is common, and the cure usually simple, since many kinds of absorbent resilient-mounting sleeves are available. In heavy units, the residual vibration of the plant may have to be separated from transmission to associated structures, in which case the whole plinth is insulated with suitable material from contact with its surrounds.

Cases have been noted of pumps mounted on steel structures which, although structurally sound from a weight-carrying point of view, are quite inadequate when stiffness is considered. Such structures should always be properly braced; particular attention should be paid to this when mounting a pump on an existing structure which is otherwise adequate.

SECTION EIGHT

TESTING AND COMMISSIONING

(a) Testing requirements

The purchaser of pumping plant generally requires the manufacturer to prove that the plant will meet the requirements which he laid down in the original order. This is achieved by the manufacturer carrying out an acceptance performance test on the completed pump, either in his works or on site, before the purchaser will officially accept the equipment. This test will prove to the purchaser that the pumping plant is working, and that it meets the requirements of the specification.

When writing the pump specification, the purchaser must indicate the tests he requires, the degree of test accuracy necessary and the form the tests shall take. He may require the plant to be tested in accordance with B.S. 599: *Pump tests*. This lays down the form that tests should take, the measurements that should be made and also specifies degrees of test accuracy, viz. Class A, B and C in descending order of accuracy. Clearly, testing in any form will be an expense on the contract concerned, and the degree of test should therefore be carefully considered, as there is little virtue in specifying a higher degree of test than is called for by the nature of the work.

Performance tests are probably the most important tests the purchaser of pumping plant need specify. When ordering a pump, he gives the manufacturer the performance he requires, and generally gives a test tolerance on this performance. For example, the usual tolerance on efficiency at duty point is $\pm 2\frac{1}{2}\%$, although tolerances are generally left to the discretion of the purchaser. The tolerance is imposed to allow for design and testing errors. If the performance falls outside these tolerances on test, the purchaser can refuse to accept the equipment, or withhold payment until the deficiencies are rectified. It is important in this respect, when ordering a pump, to allow sufficient time to be allocated, during the scheduling stage, for any modifications to the pump and for any retesting.

Performance tests should provide sufficiently accurate data to calculate and plot characteristics, including details of head, power absorbed, efficiency and flow, and to indicate the specified duty point or points. They will also indicate whether the shape of the head-quantity curve is satisfactory; for example, that it is continuously falling. It is also usual to specify the loads at which test readings shall be taken.

It is customary for manufacturers to test oil pumps with water, but the specification usually asks for the results and curves to be given for the working fluid. It is thus necessary for the manufacturer to extrapolate the test readings for water to those applicable to the fluid concerned. A similar procedure will have to be adopted if a pump is tested at a temperature below its normal working temperature.

In cases where the suction conditions are likely to be critical, i.e. where it is difficult to arrange sufficient suction head, it may be desirable to specify that the manufacturer should carry out an NPSH test in order to determine the suction characteristics of the pump.

It is usual to specify an hydraulic pressure test on the pump casing. The test pressure is normally 1·5 times the maximum working pressure.

It is sometimes desirable, in pumps operating under arduous conditions, to ask the manufacturer to carry out materials and weld testing. For pumps operating under high pressures the casing welds should be radiographically inspected. Inspection of the impellers for flaws that may initiate erosion or corrosion may be desirable. Radiographic examination of these impellers can also be carried out. The porosity of casings can be determined by carrying out a gas pressure test. This is done by pressurising the casing/vessel with anhydrous ammonia to about 1 bar and then increasing the pressure to about 4 bar with air. A sulphur dioxide source (a siphon bottle with a 3 mm bore tube) is then passed over the outside of the casing, welds, etc. If there is any porosity, it will be shown by a cloud of white ammonium sulphite fumes, a deposit of which will pin-point the porosity or leak.

Besides checking the performance of a pump, an acceptance test will also determine whether the basic mechanical design of the pump is sound. For example, it will show up faults with the internal clearances, in glands or mechanical seals and in the driver and coupling. A visual examination of the pump can also be made.

When tests are being carried out by the manufacturer, it is usual for a representative of the buyer to be present to witness the work being carried out and so ensure that the specification is complied with.

In the case of small off-the-shelf pumps, it may not be necessary for tests to be carried out. The manufacturer's stated performance can be accepted.

(b) Measuring equipment

During a works test it is necessary to measure these parameters:

(i) *Inlet and outlet head*

The total inlet and outlet heads are determined from the sum of the pressure head, the positional head of the pressure head tapping, and the velocity head calculated assuming uniform velocity in the pipe. The pressure heads are measured from static pressure tappings situated at a distance of 2 pipe-diameters from the inlet and discharge flanges. If two or more tappings are provided, it is usual to connect these through shut-off cocks to a ring manifold of cross-sectional area not less than the sum of the cross-sectional areas of the tappings. The greater the number of tappings provided, the more accurate will be the pressure readings obtained, although it is not usual to have more than four.

The pressure can be read by means of a Bourdon gauge, a U-tube or a single-leg manometer. When the heads to be measured exceed 3 m or so, water columns may be inconveniently high. Mercury manometers are suitable for heads up to about 15 m

If one leg of a U-tube manometer is connected to the pump suction branch and the other leg to the delivery branch, the gauge will give a direct indication of the head generated across the pump.

If a high degree of accuracy is required, when the velocity head is greater than 5% of the pump total head, it is usual to make arrangements to traverse the pipe on two diameters. This is achieved by means of a probe of the yaw-meter type, so that the full effect of swirl in the pipe can be taken into account.

(ii) *Discharge flow rate*

The flow rate can be measured by any of the following means:

1. *Weighing tank*

   This consists of weighing the quantity of water which flows into a reservoir in a given period of time. The temperature of the liquid must be known in order to ascertain the specific gravity. This method is subject to the errors of weighing and of the time-measuring apparatus employed, and is generally used only for measuring small discharges.

2. *Volumetric tank*

   This method involves measuring the time taken for a given volume of water to enter a reservoir. It is usual to arrange for a change-over device on the outlet of the delivery pipe, which permits the flow to be directed into the measuring tank when flow conditions are steady, and away from it when measurement has been completed. The method is subject to the same kind of errors as the weighing tank method.

3. *Uncalibrated orifice plates, venturi meters and nozzles*

   The measurement of discharge can be carried out using a square-edged orifice plate, or a venturi meter or nozzle fitted in a straight run of pipeline, which should be of circular cross-section. The pipe should be full of water on both sides of the instrument, and the two pressure tappings on either side of the instrument connected to a differential manometer. B.S. 1042: *Code for flow measurement*, deals with flow measurement in more detail.

4. *Uncalibrated notches and weirs*

   The flow rate can also be determined by discharging the pump flow over a notch or weir.

5. *Current meters*

   These are generally used for measuring the flow of large pumps when tested at site. The method consists of a number of small generators driven by impellers, which are fixed to a grid in the discharge pipe or culvert. The water flow drives the impellers; the current developed by each generator is measured and can be related directly to velocity. The discharge is obtained by integrating the velocity over the area of the cross-section of the pipe.

6. *Radioactive dilution*

This method is used in the main for the site testing of large circulating water pumps. The method involves the measurement of flow, based on a comparison between the concentration of a tracer, continuously introduced at a known rate into the system, with the concentration of samples removed from some position, sufficiently distant from the injection point to ensure thorough mixing. If conducted with sufficient care, the method can give quite accurate results.

A variation on the above is the radioactive velocity method. This involves measuring the time taken for a radioactive sample to pass between two points in the system.

7. *Heat-balance*

The flow rate in a condenser cooling water system can be determined by carrying out a heat-balance across the condenser. Accurate measurement of temperatures is necessary to ensure accuracy in the value of flow obtained.

(iii) *Measurement of speed*

Although a revolution counter, in conjunction with a stop-watch, will usually give a more reliable indication of rotational speed than a tachometer, the latter may often be essential in order to verify that the speed is held constant during the test run. If a tachometer only is used, it should be accurately calibrated. In electric tachometers, the indicating instrument is in effect a voltmeter actuated by the current generated by a small dynamo coupled to the pump shaft.

If the pump is driven by a synchronous or induction motor, taking its supply from a frequency-controlled network, the shaft speed can be accurately determined by measuring the supply frequency and the slip frequency.

(iv) *Measurement of power*

The pump input power can be determined by any of the following methods:

1. *Measuring the electrical power to the driving motor.*

If the motor is supplied by a 3-phase A.C. supply, the two watt-motor method of power measurement is usually used. If by a D.C. supply, power is obtained by measuring voltage and current. For both of these methods it is necessary to know accurately the efficiency of the driving motor.

2. *Measuring the torque*

This can be achieved by using a dynamo-motor (such as a swinging-yoke motor) wherein the driving motor is of special construction having its yoke or frame mounted on ball bearings co-axial with the shaft bearings. As soon as the motor is energised, the frame tends to revolve in a direction contrary to that of the armature or rotor, a tendency that is counteracted either by weights suspended on an arm fixed to the motor frame, or by a spring-balance attached to the end of this arm. In this way, a reading of torque can be calculated.

The torque being transmitted by the driver can also be measured directly by a torquemeter coupled between the driver and the pump.

Temperature of the working fluid can be determined by ordinary mercury thermometers in suitable pockets, by thermocouples or capillary thermometers. B.S. 2765: *Dimensioning system and terminology for industrial temperature-detecting elements and pockets*, deals with thermometer pockets.

(c) Special test rigs

(i) *Open-circuit test-rigs*

Fig. 8.1 shows a simple test-rig with a measuring tank, for the works testing of a pump set. The pump, bolted to the test plate which forms the cover of the underground reservoir, draws water through a short suction pipe and discharges it through the throttle

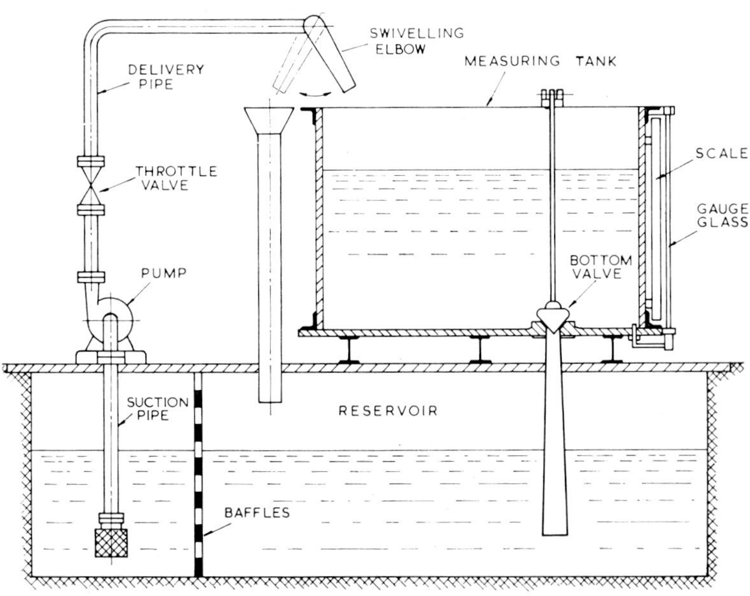

Fig. 8.1 Simple test-rig with measuring tank.

valve to a swivelling elbow, which directs the flow into either the measuring tank or the reservoir. The quantity of water in the measuring tank is shown by a glass gauge-tube. An outlet pipe, provided with a streamlined valve, permits the tank to be emptied quickly. Baffles in the sump ensure that the pump draws from reasonably still water. A tank mounted on a weighing machine would be just as satisfactory as the volumetric tank shown in the sketch.

A possible source of error results from the fall in level in the suction reservoir as the measuring tank fills. If the pump speed and throttle valve positions are held constant, variation in the suction lift will alter the total head on the pump, and the discharge will be affected correspondingly. If the sump is large and/or the pump head high, any variation in the suction lift can be ignored, but in some cases it could considerably affect the test result. This can be partly avoided by taking instantaneous readings of suction and delivery head exactly half-way through the run. This assumes that the discharge attains its mean value half-way through the run.

Fig. 8.2 shows in diagrammatic form the layout of a test-rig with a measuring weir. This arrangement is generally suitable only for low flows, and inaccuracy in reading the head of water at the weir may be a problem.

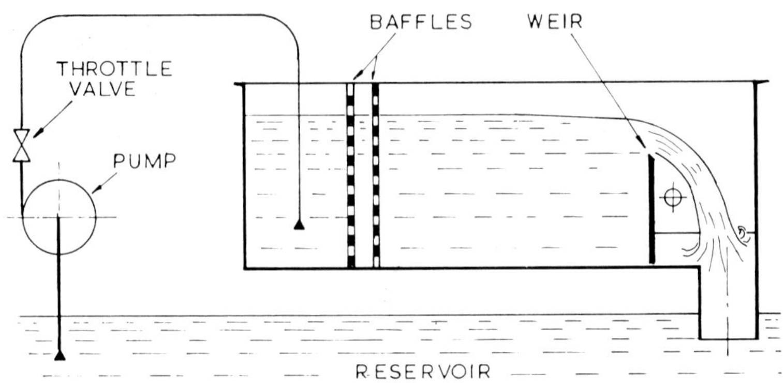

Fig. 8.2 Diagrammatic layout of an open-circuit test-rig with measuring weir.

Fig. 8.3 shows the layout of a test-rig using venturi meters for measurement of flow. Two venturies are generally used, so that for measuring of low flows, one can be used to obtain greater accuracy.

(ii) *Closed-circuit test-rigs*

These are used to test pumps of large capacity which would require very large measuring reservoirs if they were tested in open-circuit systems. One of the disadvantages of a closed-circuit

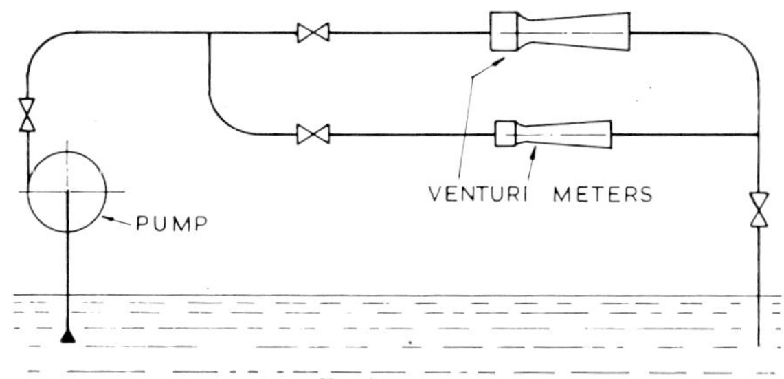

Fig. 8.3 Diagrammatic layout of an open-circuit test-rig using venturi meter for flow measurement.

system is the heat generated, which results in an increase in the water temperature.

Fig. 8.4 shows a system for the testing of boiler feed pumps. The suction head of the pump can be simulated by forcing air into the closed drum, the desired water temperature being obtained by injecting steam into the liquid.

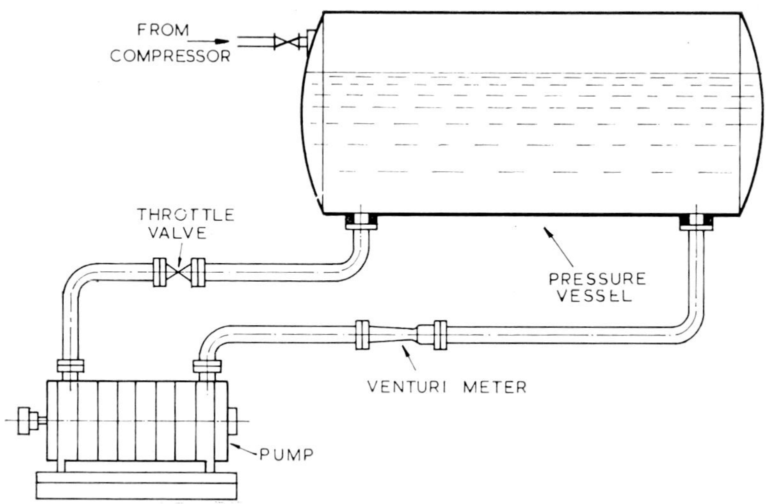

Fig. 8.4 Closed-circuit system for testing boiler feed pumps.

Fig. 8.5 shows a system for testing pumps with high suction heads.

(iii) *Rig for cavitation testing*

Fig. 8.6 shows the layout of a test rig for investigating cavitation. This is carried out by decreasing the suction head by throttling a valve on pump suction while maintaining the discharge constant. The valve should be located at such a distance from the pump inlet that it produces no perceptible disturbance in the flow.

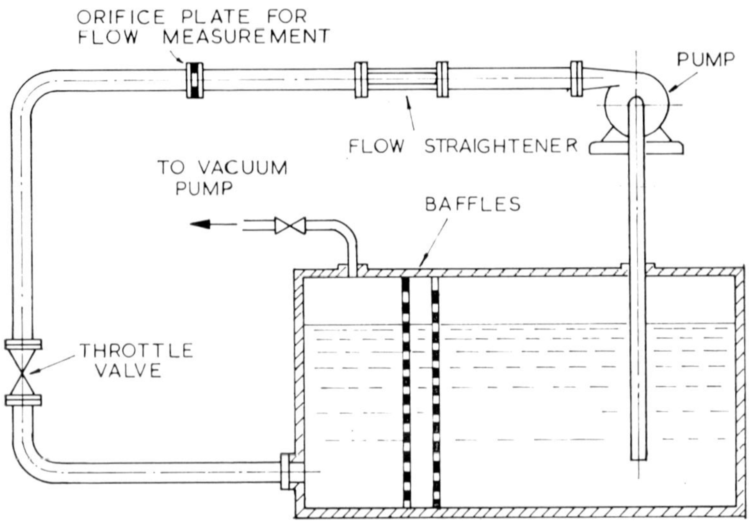

Fig. 8.5 Closed-circuit system for testing pumps operating with a high suction head.

(d) Extrapolative tests

When testing large pumps, it is possible that the manufacturer cannot meet the power requirement necessary to test the pump at full-load. This is generally due to a restriction imposed, by the local Electricity Board, on the electrical load that can be taken by the works.

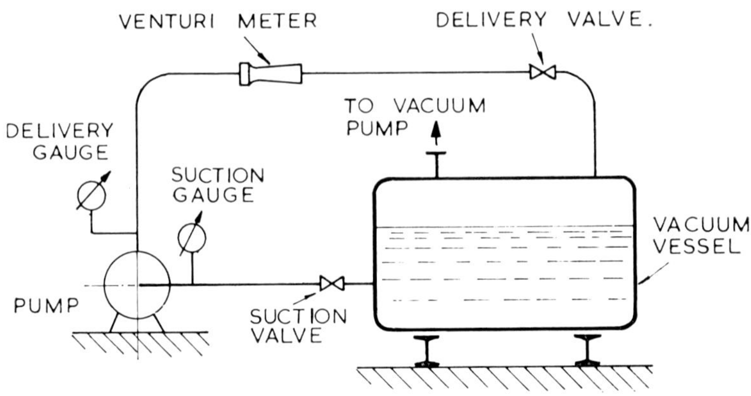

Fig. 8.6 Diagrammatic layout of a test-rig for cavitation tests.

In such cases, it is usual to test the pump at reduced speed, and to extrapolate the values of head and discharge obtained to those which would be obtained at full rated speed. This is achieved as follows:

Let N_1, H_1 and Q_1 be the speed, head and discharge at the test condition.

N_2, H_2 and Q_2 be the speed, head and discharge at the rated pump conditions.

For a centrifugal pump, H is proportional to N^2.

$\therefore \quad H_1 = KN_1^2$

and $H_2 = KN_2^2$ and where K=constant

$$\therefore \quad \frac{H_1}{N_1^2} = \frac{H_2}{N_2^2}$$

and $H_2 = H_1 \left(\frac{N_2}{N_1}\right)^2$

Similarly, for a centrifugal pump, Q is proportion to N, and

$$Q_2 = Q_1 \frac{N_2}{N_1}$$

power is proportional N^3

$\therefore$ power at rated speed = power at test speed $\times \left(\frac{N_2}{N_1}\right)^3$

The cube of the speed is based on the assumption that efficiency remains unchanged with speed.

Having plotted the characteristic curves for the test speed, the curves for the full rated speed condition can be obtained by applying these formulae. This is shown in Fig. 8.7.

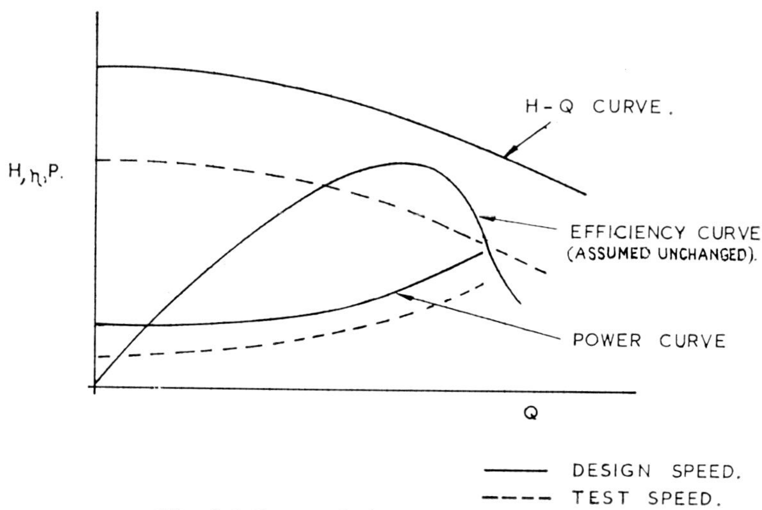

Fig. 8.7 Extrapolative test characteristics

(e) Site testing and commissioning

Performance tests may have to be carried out on-site for either of (or both) the following reasons:

- The pump may be physically too large, or the power requirement too heavy, for it to be run in the maker's works.
- Periodical tests may be necessary during the life of the pump to check on its performance. This may be particularly so after mainten-

ance periods, or after the pump has been reassembled following repair.

The main problem in carrying out a site test is generally that of arranging instrumentation of a suitable type and accuracy. The instruments normally provided for use during general running are invariably of the wrong type and insufficiently accurate, so that special measuring equipment must usually be transported to the site.

Head and speed measurements can be made in the same manner as for a works test, although in some cases the accuracy of pressure measurement may be affected by the pump not having straight lengths of pipe before the inlet and after the discharge. Power can be measured by the wattmeter method or by a torquemeter. In the latter case, the pump unit must be designed with this in mind, so that sufficient space is left between the pump and driver to receive the meter. After testing, this is removed and replaced by a jack shaft.

The most difficult parameter to measure at site is normally the flow, since there is often no room (or no straight section of pipe in the system) to allow the installation of an orifice plate or venturi meter. If this is the case, the current meters installed either in the suction or discharge pipe can be used. Considerable development has been carried out recently on the radioactive dilution method of flow measurement, which particularly lends itself to the measurement of flow rate during a site test: it has been proved to give results of considerable accuracy.

Whenever possible, in designing a system that is to be tested at site, consideration should be given to the installation of a straight section of pipe that is capable of receiving an orifice plate or venturi meter, which can be replaced by a normal pipe length when the test is completed.

On the whole, site tests are more expensive, more difficult to carry out and less accurate than works tests, although careful consideration and thought given to the testing of the system during the design stage can eliminate many of the possible errors and difficulties.

When commissioning a pump set, the following are precautions that should be taken:

(i) Check instruction manual.

(ii) Care should be taken during construction of the system to ensure cleanliness and freedom from debris, such as rags, cotton waste, iron filings and pieces of wood or metal, etc.

(iii) Thorough checks should be made before start-up to ensure that bearings are filled with lubricant, that the pump is fully primed and the system filled with water.

(iv) The pump should initially be run at low load, to ensure that the system is working satisfactorily.

(v) During the initial running hours of the pump, a careful check should be made to ensure that bearings and glands are not overheating, and that there is no excessive leakage from glands or joints.

(vi) After a period of running, the holding-down bolts and couplings etc. should be checked for tightness.

(f) Spares

The number and extent of spares carried depends on the amount of stand-by plant available, and the likely reliability of the plant. The economic aspect must also be considered.

In many cases, it may prove an advantage to have a complete spare pump in the stores which can quickly be changed over with the faulty pump.

As a general rule, the following is a list of spares that should normally be carried. The list is only a guide and certain specific applications may require individual assessment to ascertain the complement of spares for one pump.

(i) *Pump*

1 complete rotating assembly.
1 set of impeller wearing rings and bushes.
1 set of bearings.
1 thrust bearing.
Gland packing, or, in the case of pumps with mechanical seals, spare sealing rings and seats.

(ii) *Motor*

1 set of bearings.
1 set of exciter brushes (if synchronous motor).
1 set of brush holders (if synchronous motor).
1 set of exciter bearings.

It is sometimes desirable to specify short-term and long-term spares. That is, the manufacturer should supply with the pump spares that may be required during the early life of the pump, say, the first year or so, or during the guarantee period. Long-term spares are those that will be required after this period, and generally consist of the large items, like shafts, impellers, casings, etc.

SECTION NINE

STANDARDISATION AND SPECIFICATION

(a) Standardisation

It will be apparent that, in very many instances, particularly with process pumps, there is a considerable case for the standardisation of both quality and dimensions. It is of great convenience, for instance, to ensure that a pump bought for a certain range of duties is dimensionally identical from one manufacturer to the other, and on such a basis that interchangeability is available from any manufacturer on replacement items.

In specification, the work of defining the article needed is greatly simplified, and in quotation of the relevant standard a reliable minimum is readily obtained. Equally, the manufacturer has a basis of design which rationalises the multiplicity of equipment which he would otherwise be called upon to supply, and is able, as a result, to concentrate his efforts with beneficial results to delivery dates and prices.

Standardisation has, of course, been one of dimensions, leaving scope for impeller detail design and materials to the manufacturers concerned. The user may therefore look for the best design available in a certain dimensional range, and the British Standards Institution covers the usual process pump requirements with the following standards:

B.S. 599 *Methods of testing pumps*

B.S. 3829 Centrifugal pumps. Part 1 *Horizontally split casing double inlet impeller.*

B.S. 3829 Centrifugal pumps. Part 2 *End suction pumps.*

B.S. 4082 *Vertical in-line centrifugal pumps.*

B.S. 4519 Part 1 *Horizontal end-suction centrifugal pumps.* (To be revised to ISO requirements).

In each case, the B.S. covers the principal external dimensions aimed at the mechanical interchangeability of single-stage centrifugal pumps of similar speed. Dimensions are given for flange position of inlet and outlet, shaft diameters and shaft heights for coupling connections. The capability of any manufacturer's design to fit into pre-designed pipework systems will be readily appreciated.

Much more could be done, however, and it has not yet been found neces- to extend the range of standardisation into freely connected pumps, multi-stage pumps or pumps for the larger duties.

The following British Standards do, however, cover pumps for small duties which the reader may have occasion to use:

B.S. 1394 *Power driven circulators for heating plants.* Directed coupled pumps for heating mains with water not exceeding 93·5° C in temperature, and static heads not exceeding 30 m. This

Table 9.1 A Pump Specification Sheet

	Data/requisition sheet for ROTODYNAMIC PUMPS			No. :		page
				Job No.:		
				No.:		
1	PUMP NUMBER : P —			Number required:	(incl.	spare)
2	OPERATING DATA					
3	Liquid :			Specific gravity at pumping temperature:		
4	Service : intermittent/continuous			Viscosity at pumping temperature :		10^{-6} m²/s
5	Pumping temperature :	°C		Vapour press. (abs.) at pumping temp. :		kg/cm² abs
6	Capacity, normal/maximum : /	m³/h		Absolute press. in suction vessel :		kg/cm² abs
7				Absolute press. in discharge vessel :		kg/cm² abs
8	PUMPING DATA					
9	Absolute pressure in suction vessel :	m liq abs		Absolute pressure in discharge vessel :		m liq abs
10	Height of liquid level in suction vessel			Height of liquid level in discharge vessel		
11	above/below pump centre line (min.) :	m liq		above/below pump centre line (max.) :		m liq
12	Total pressure loss in suction line			Total pressure loss in discharge line		
13	between suction vessel and pump :	m liq		between pump and discharge vessel :		m liq
14	Total minimum suction head :	m liq abs		Total maximum discharge head :		m liq abs
15	Vapour pressure (absolute) of liquid :	m liq abs		Differential head at maximum capacity :		m liq
16	Minimum available N P S H :	m liq		Estimated normal power required :		watts
17	MECHANICAL DATA					
18	TYPE:			Speed :		rpm
19				Mechanical seal : yes/no		
20	Direction of rot. (facing driven end): cw/ccw (preferably ccw)			Circulation connection : yes/no		
21	Cooling required on : stuffing box/bearings/pedestal			Smothering gland : yes/no		
22	Cooling water : fresh/brackish/salt			Lantern ring conn. on stuffing box : yes/no		
23	Cooling water, max. temp. at inlet : °C at	kg/cm² ga		Type of bearing lubrication : grease/forced/ring oiled		
24	Position of nozzles (facing driven end) and flange standards					
25	Suction: end/top/side, BS/ASA	Discharge: end/top/side, BS/ASA		U-turn/In-line		

MATERIAL SPECIFICATION		ACCESSORIES		
Parts	spec. No.	Base plate	:	combined/separate
Casing and cover		Foundation bolts	:	yes/no
Impeller		Coupling	:	flexible/all-metal/spacer type
Impeller wearing ring		Coupling guard	:	yes/no non-sparking
Casing wearing ring		Pressure/suction gauges	:	yes/no
Shaft		Cooling water harness	:	yes/no open/closed system
Shaft sleeve		Seal harness		discharge to seal/seal to suction/
Gland				strainer/flow controller
Gaskets to be alkaline-resistant	yes/no			
Mechanical seal parts				
Cooling water piping				

PUMP DRIVE	REFERENCES
Electric motor (cycles 50–60)/steam turbine/comb. engine	For inspection, testing, preparation of test data and curves and
Drive : direct/gear/V-belt/variable speed	for information to be submitted with the tender refer
Fstimated installed power : watts	to contract documents

REMARKS AND/OR DESCRIPTION OF REVISIONS

Made by	Date	EQUIPM. :	Rev. letter				
Checked by	Date	PLANT :	Date				
			Sign.				
Appr. by	Date	CONSIGNEE:	Sheet No. 1	cont'd on sheet No.			
Eng. by :			Req. No.				
Principal:							

standard specifies quality of material, alignment, assembly, installation and connections.

B.S. 2896 *Built-on coolant pumps for internal combustion engines (nominal output 1750–17500 gal/h at 30-ft maximum head).* Covers dimensions generally governing interchangeability.

B.S. 2993 *Coolant pumps for internal combustion engines.* Gives housing dimensions for seals, shaft diameters and stubs.

B.S. 3766 *Machine tool motorised centrifugal type coolant pumps.* Gives *Q–H* performance characteristics for ten sizes of pumps.

(b) Specification

The reader will realise that the above-mentioned standardisation will generally cover only physical dimensions, and still leaves very many areas of specification to be covered in detail order sheets. Much time can be lost in the preparation of tenders, and quite basic areas of design may be overlooked unless a comprehensive 'duty specification' is drawn up at the time the manufacturer is asked to tender for the pump required.

Equally, it is important that all manufacturers tender on a similar basis, if true tenders are to be received, each fully covering the user's particular requirements. Clarity and comprehensiveness is of the utmost importance to both.

With the intention, therefore, that the scope of tendering be attended to as completely as possible, the list of requirements (Table 9.1) is given It covers all normal requirements, and at the same time acts as an aide-memoire to the user in ensuring that the remainder of the system is carefully considered, before the pump specification is made out.

APPENDIX A

MATERIALS OF CONSTRUCTION

Legend: SS = Stainless steel
CI = Cast iron
CS = Cast steel
S.G. = Specific gravity

Fluid, specific gravity, and viscosity (centipoise) at 20deg.C (68deg.F)	Material for			Remarks
	Casing	Shaft	Impeller	
Acetaldehyde S.G. 0·78 Visc. ·23	SS B.S. 1504–821 or 845	SS B.S. 1501–821 or 845	SS B.S. 1504–821 or 845	B.S. 1504–845 more suitable for higher temperatures
Acetic acid S.G. 1·05 Visc. 1·22	SS B.S. 1504–821 or 845	SS B.S. 1501–821 or 845	SS B.S. 1504–821 or 845	B.S. 1504–845 more suitable for higher temperatures
Acetic anhydride S.G. 1·08 Visc. 0·91	SS B.S. 1504–821 or 845	SS B.S. 1501–821 or 845	SS B.S. 1504–821 or 845	B.S. 1504–845 more suitable for higher temperatures
Acetone S.G. 0·79 Visc. 0·33	CI	SS B.S. 1501–821	CI	
Aliphatic Amines (Di Methyl and and Di Ethyl)	CI or SS B.S. 1504–821 or 845	SS B.S. 1501–821 or 845	CI or SS B.S. 1504–821 or 845	
Ammonia S.G. ·61 Visc. ·235 at 0 deg. C	CI or CS	SS B.S. 1501–821	CI or SS SS B.S. 1504–821	
Ammonium hydroxide	CI	SS B.S. 1501–821	CI	
Ammonium sulphate solution				
Benzene S.G. ·88 Visc. ·65	CI	SS B.S. 1501–821	CI	
Butanol S.G. ·81 Visc. 2·95	CI	SS B.S. 1501–821	CI or SS B.S. 1504–821	
Butyl acetate	CI	SS B.S. 1501–821	CI	
Calcium chloride brine	CI	Aluminium Bronze	CI	

Fluid, specific gravity, and viscosity (centipoise) at 20deg.C (68deg.F)	Material for			Remarks
	Casing	Shaft	Impeller	
Calcium hypochlorite (aqueous solution. Bleach liquor)				
Carbon disulphide				
Carbon tetrachloride				
Copper sulphate solution				
Cresol	CI	CS	CI	
Diketene S.G. 1·09 Visc. 1·02	SS B.S. 1504–821	SS B.S. 1501–821	SS B.S. 1504–821	
Dowtherm	CI or CS	CS or SS B.S. 1501–821	CI or CS	Carbon steel to be used if temperature over 100deg.C. Dowtherm E sometimes breaks down to HCl in which case SS must be used throughout.
Ethanol S.G. 0·79 Visc. 1·2	CI	SS B.S. 1501–821	CI	
Ethyl acetate	CI	SS B.S. 1501–821	CI or SS En 588	
Ethyl hexanol	CI	SS B.S. 1501–821	CI	
Ethylene glycol	CI	SS B.S. 1501–821	CI	
Ethylene oxide	CI	CS	CI	
Fatty acids (including lactic, oleic and stearic acids)	SS B.S. 1504–821 or 845	SS B.S. 1501–821 or 845	SS B.S. 1504–821 or 845	
Ferric chloride (aqueous solution)				

Fluid specific gravity and viscosity (centipoise) at 20deg.C (68deg.F)	Material for			Remarks
	Casing	Shaft	Impeller	
Ferrous sulphate (aqueous solution)				
Ferrous sulphate solution				
Formaldehyde (formalin solution) S.G. 1·08 Visc. 2·03 at 25degC	Gunmetal	SS B.S. 1501–821	SS B.S. 1501–821	
Formic acid	SS B.S. 1504–845 Ti	SS B.S. 1504–845 Ti	SS B.S. 1504–845 Ti	
Furfural	CI	SS B.S. 1501–821	SS B.S. 1504–821	
Gin	SS B.S. 1504–821	SS B.S. 1501–821	SS B.S. 1504–821	
Gasoline (petrol)	CI	SS B.S. 1501–821	CI	
PVC Slurry	SS B.S. 1504–821	SS B.S. 1501–821	SS B.S. 1504–821	
Glycerol	CI or gunmetal	SS B.S. 1501–821	CI or gunmetal	
Heat transfer oils (e.g. Shell volute 45 and Mobiltherm)	CI or CS	SS B.S. 1501–821	CI or CS	
Hydrocarbons (aliphatic or aromatic)	CI	SS B.S. 1501–821	CI	
Hydrochloric acid	Ceramic or rubber lined Hasteloy B	Ceramic or rubber lined Hasteloy B	Ceramic or rubber lined Hasteloy B	
Hydrogen peroxide	99·8% Aluminium or SS B.S. 1504–821	SS B.S. 1501–821	99·8% Aluminium or SS B.S. 1504–821	Aluminium impeller to be of the open-type

Fluid specific gravity and viscosity (centipoise) at 20deg.C (68deg.F)	Material of			Remarks
	Casing	Shaft	Impeller	
I.M.S.	CI	SS B.S. 1501–821	SS B.S. 1504–821	
Indene	CI	SS B.S. 1501–821	CI	
Isopropanol	CI	SS B.S. 1501–821	CI	
Methanol S.G. 0·79 Visc. 0·6	CI	SS B.S. 1501–821	CI or SS B.S. 1504–821	
Molasses†	CI*	CS or SS B.S. 1501–821	CI* or Bronze	*at ambient temperature only †positive displacement pumps may prove more suitable than rotodynamic
Nitric Acid	SS B.S. 1504–821	SS B.S. 1501–821	SS B.S. 1504–821	20°C all concentrations
Nitrogen	CI	CS	CI	
Nitrous acid				
Oxalic acid				
Paraldehyde	SS B.S. 1504–821	SS B.S. 1501–821	SS B.S. 1504–821	
Phosphoric acid	SS B.S. 1504–845	SS B.S. 1504–845	SS B.S. 1504–845	
Phthalates	CI	SS B.S. 1501–821	CI	All grades, higher and lower
Plasticiser	SS B.S. 1504–821	SS B.S. 1501–821	SS B.S. 1504–821	
Polyvinyl alcohol	SS B.S. 1504–821	SS B.S. 1501–821	SS B.S. 1504–821	
PVA Emulsions	SS B.S. 1504–821	SS B.S. 1501–821	SS B.S. 1504–821	

Fluid specific gravity and viscosity (centipoise) at 20deg,C (68deg,F)	Material of			Remarks
	Casing	Shaft	Impeller	
Sea water	Bronze or ductile iron 20% Cr, 2% Ni B.S. 970 En 58 (b)	Monel or steel protected	Bronze or ductile iron 20% Cr, 2% Ni B.S. 970 En 58 (b)	
Sewage (not effluent)	CI	SS B.S. 1501–821	CI	
Soap				
Soda ash solution	CI	SS B.S. 1501–821	CI	
Sodium bisulphite	Ceramic or CI, lead or titanium lined	Ceramic or CS, lead or titanium lined	Ceramic or CS, lead or titanium lined	
Sodium chloride brine	CI rubber lined	SS B.S. 1501–821	CI rubber lined	
Sodium hydroxide (caustic soda)	CI	SS B.S. 1501–821	CI	Solutions
Sodium hypochlorite	CI rubber or Penton lined	CI rubber or Penton lined	CI rubber or Penton lined	
Sodium nitrate brine	CI	CI	CI	
Sodium thio-sulphite				
SPB 3 and SPB 6	CI	SS B.S. 1501–821	SS B.S. 1504–821	
Styrene	CI	SS B.S. 1501–821	SS B.S. 1504–821	
Sulphuric acid (96% and over) S.G. 1·84 Visc. 25	CI	SS B.S. 1501–821	CI	

Fluid specific gravity and viscosity (centipoise) at 20deg.C (68deg.F)	Material of			Remarks
	Casing	Shaft	Impeller	
Sulphuric acid (below 96%)	Ceramic or cast Si-iron 14·5% Si Hasteloy D	Ceramic or Hastelloy B steel protected	Ceramic or cast Si-iron 14·5% Si Hasteloy D	
Sulphurous acid				
Tar				
Toluene S.G. 0·87 Visc. 0·59	SS B.S. 1504–821	SS B.S. 1501–821	SS B.S. 1504–821	Carbon steel can be used, but will discolour product
Triacetin S.G. 1·16 Visc. 23·45	CI	SS B.S. 1501–821	CI or bronze	
Trichloroethane				
Trichloroethylene	CI	SS B.S. 1501–821	CI	
Vinyl acetate	CI	SS B.S. 1501–821	CI	Copper bearing materials must not be used
Vinyl chloride	SS B.S. 1504–821	SS B.S. 1501–821	SS B.S. 1504–821	
Water	CI or bronze	SS B.S. 1501–821, mild steel or bronze	Bronze, CI or gunmetal	
Water with PVC in suspension	SS B.S. 1504–821	SS B.S. 1501–821	SS B.S. 1504–821	
Water (deionised)	CI or SS B.S. 1504–821	Mild steel, SS B.S.1501–821, bronze	CI or SS B.S. 1504–821	
White spirit	CI	SS B.S. 1501–821	SS B.S. 1504–821	
Whisky (blended, matured)	SS B.S. 1504–821	SS B.S. 1501–821	SS B.S. 1504–821	

Fluid specific gravity and viscosity (centipoise) at 20deg.C (68deg.F)	Material for			Remarks
	Casing	Shaft	Impeller	
Whisky spirit	Bronze or SS B.S. 1504–821	SS B.S. 1501–821	Bronze or SS B.S. 1504–821	
Xylene	SS B.S. 1504–821	SS B.S. 1501–821	CI	
Yeast suspension	SS B.S. 1504–821	SS B.S. 1501–821	SS B.S. 1504–821	

APPENDIX B:

COUPLING TYPES AND USAGE

Type	Power* transmitted	Accessibility	Alignment (ease of checking with coupling in position)	Permissible mis-alignment* and resilience	Effect of failure of flexible member or members	Speed* range	Lubrication	Environmental limitations	Remarks
Barrel	Fractional and low power drives	As coupling cannot be split, driving and driven shafts have to be parted for assembly and removal	Difficult to check alignment on some designs	Considerable angular and parallel mis-alignment can be accommodated. End-float limited to resilience of material. Very flexible in torsion, damping out torsional vibrations and absorbing shock loads	Drive between shafts will immediately cease	Normally up to 5000 rpm. Higher speeds permiss-ible, but at reduced load capacity	None required	Resilient material has to be suitable for environment. Temperature range with standard material 0 to 65°C	Most suited to very light fractional hp drives, in which range this type of coupling is very economical
Chain	Normal range 4–750 kW at 1000 rpm	Readily split in position. All parts removed radially, so no additional end clearance required	Difficult to check parallel mis-alignment. Angular align-ment easily checked with chain removed	Angular up to 1 deg. Parallel 0·25–0·75 mm. End-float limited to clearance between sprocket teeth and chain side links. Not flexible in torsion	If chain breaks, drive between shafts will probably be retained, but may cease immediately. In the unlikely event of sprocket teeth shearing, drive will cease immediately	1000–6000 rpm	Requires lubrication	Little affected by enrivonment. Special lubrication will be required for high temperature applications	Some inherent backlash, so not suitable for cyclic or reversing loads
Disc: metallic	Normal range from fractional to 375 kW at 1000 rpm	Readily split in position. Some end clearance may be required to remove the bolts. Spacer types enable easy removal of one unit without disturbing the other	Alignment easily checked	Angular up to 1½ deg. Parallel, very limited, particularly for single disc designs. End float between 1·5–3·0 mm. Not flexible in torsion	Drive between shafts will cease immed-iately, except for some single-disc designs, where drive may be main-tained by interference of the bolts clamping the disc	Suitable for very high speed applica-tions	None required	Little affected by environment	Requires very little maintenance, but has to be stripped to examine thoroughly. Initial cost may be comparatively high

* *Note:* Unless otherwise stated, figures given are typical only of the normal range of standard production items available. The precise value for any of these characteristics, within the broad ranges given, depends on the size and design of the coupling, and can only be obtained from the manufacturer.

Appendix B: Coupling Types and Usage (continued)

Type	Power* transmitted	Accessibility	Alignment (ease of checking with coupling in position)	Permissible mis-alignment* and resilience	Effect of failure of flexible member or members	Speed* range	Lubrication	Environmental limitations	Remarks
Disc: non-metallic	Normal range 2·25–6 MW at 1000 rpm (but see note under Remarks)	Designs vary widely. Driving pins and disc usually removable in position, but will require some end clearance. Some designs require shafts to be parted axially before disc can be removed. Bonded rubber discs usually more easily removed	Parallel alignment easily checked, but angular alignment difficult with disc in position	Angular up to 2 deg. for rubber-impregnated fabric disc and up to 5 deg. for bonded rubber disc. Parallel mis-alignment limited to resilience of the disc. Restricted end-float, particularly for bonded rubber disc. Flexible in torsion, damping out torsional vibrations and absorbing shock loads, particularly for bonded-rubber discs	With rubber-impregnated fabric disc, drive will be maintained by interference between the driving and driven pins. With bonded-rubber disc, drive will cease immediately	Normally 1100–6000 rpm, but rubber-impregnated fabric discs not recommended for the higher speeds	None required	Resilient material has to be suitable for the environment. Temperature range of standard materials 0 to 65°C	Discs are not readily inspected in position. Designs in which the driving pins are a loose fit in the disc are not suitable for cyclic or reversing loads. Not recommended for transmitting powers greater than 75 kW per 1000 rpm
Gear	Very heavy duty in the range from 7·5 kW at 1000 rpm to 11 MW at 100 rpm	A clearance of at least 12 mm on each side of the coupling is required to slide outer gear ring off gears, to part coupling, or for inspection, aligning etc.	Angular and parallel alignment readily checked with outer gear ring removed	Angular, normally up to 1½ deg. but special designs have been made to operate up to 14 deg. Parallel from 1·5–20 mm depending on size and type. Designs incorporating only one gear can only accommodate very limited mis-alignment. End float between 1·5 and 3·0 mm. Not flexible in torsion	In the unlikely event of gear teeth shearing, the drive will cease immediately with intermittent snatching as parts of gears become jammed between housing and hub	Normally between 1000 and 100 rpm	Lubrication required	Little affected by environment. Temperature limited only by type of lubricant used	Necessary to strip coupling to inspect thoroughly for wear. If wear does develop, the whole coupling has to be replaced
Pin	Normally in the range 3 kW at 1000 rpm to 3 MW at 500 rpm	As for non-metallic disc type	Parallel and angular alignment easily checked in position	Will accommodate only very slight angular and parallel mis-alignment and end-float. Slightly flexible in torsion absorbing shock loads, but not very effective in damping out torsional vibrations	In the unlikely event of the pins shearing, the drive will cease immediately. If the resilient bushes fail, drive will be maintained by the pins	Normally between 6000 and 400 rpm	None required	Resilient material has to be suitable for the environment. Temperature range of standard materials 0 to 65°C	Designs in which the resilient bushes are a loose fit in their housing are not suitable for cyclic or reversing loads. Designs using bonded rubber bushes can be used under these conditions

**Note:* Unless otherwise stated, figures given are typical only of the normal range of standard production items available. The precise values for any of these characteristics, within the broad ranges given, depends on the size and design of the coupling, and can only be obtained from the manufacturer.

Appendix B: Coupling Types and Usage (continued)

Type	hp* transmitted	Accessibility	Alignment	Permissible mis-alignment* and resilience	Effect of failure of flexible member or members	Speed* range	Lubrication	Environmental limitations	Remarks
Spider	Suitable for fractional to 15 kW at 1000 rpm (but see Remarks column)	The majority of designs require the shafts to be parted axially for removal and insertion of spider	Parallel alignment easily checked but angular alignment difficult with spider in position	Angular up to about 1½ deg. Parallel mis-alignment and end-float limited by resilience of spider. Flexible in torsion, damping out torsional vibrations and absorbing shock loads	Drive will be maintained by interference between the projecting lugs of the two hubs	Normally between 10,000 and 4,000 rpm	None required	Resilient material has to be suitable for the environment. Temperature grane of standard material 0 to 65°C	Readily inspected for wear in position. Not recommended for applications transmitting more than 750 W per 1000 rpm
Spring	Heavy duty in the range 1·5 kW at 1000 rpm to 3·75 MW at 100 rpm	Radially split couplings require a clearance of at least 12 mm on each side of the coupling to slide back outer casing for aligning and/or connecting and disconnecting. Axially split couplings require no end clearance	Angular and parallel alignment readily checked with outer cover removed	Angular up to about 3 deg. Parallel up to 0·5 mm End-float is usually fairly limited. Flexible in torsion, damping out torsional vibrations and absorbing shock loads. Couplings can be obtained with springs and slots specially designed to de-tune the particular vibration of any application.	In the unlikely event of the springs shearing, the drive will cease immediately	Normally between 3000 and 300 rpm, but couplings suitable for higher speeds readily available	Lubrication required	Little affected by environment. Temperature limited only by type of lubricant used	Will run for long periods under arduous conditions without attention. Readily inspected for wear after removal of outer cover
Tyre	150–300 kW at 1000 rpm	Some clearance may be required to allow access to the tyre clamping screws nut, none required otherwise to remove the tyre, thus splitting the coupling	Angular and parallel alignment are fairly easily checked with tyre removed	Angular up to 4 deg. Parallel up to 3 mm End float up to 7·5 mm Flexible in torsion, damping out torsional vibrations and absorbing shock loads	Drive will cease immediately	Normally between 4500 and 1600 rpm	None required	Resilient material has to be suitable for the environment. Temperature range of standard material 0 to 65°C	Easily inspected for wear. Will run for long periods without attention. Due to easy flexing of the tyre, mis-alignment of the shafts does not cause heavy loads on the shaft bearings.

**Note:* Unless otherwise stated, figures given are typical only of the normal range of standard production items available. The precise value for any of these characteristics, within the broad ranges given, depends on the size and design of the coupling, and can only be obtained from the manufacturer.

TYPICAL DRIVE SHAFT COUPLINGS

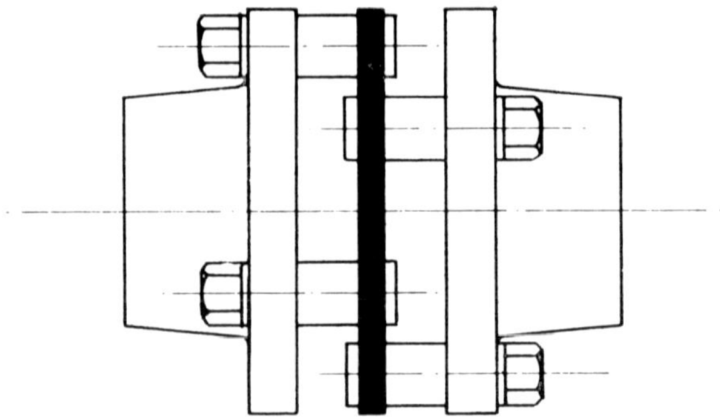

Single disc

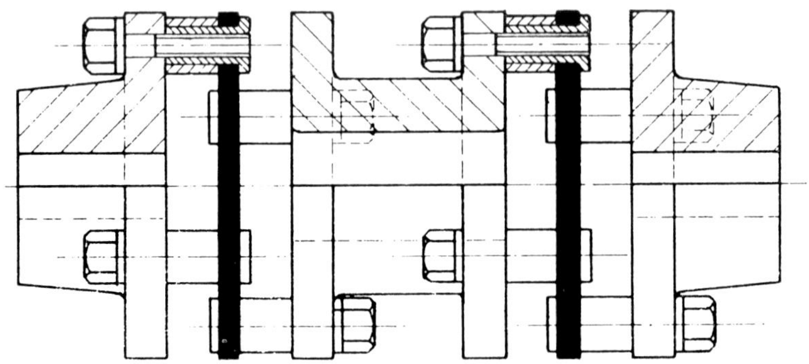

Two single discs with spacer

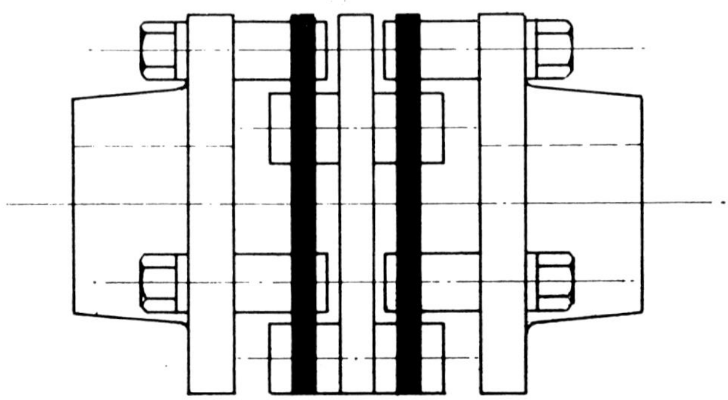

Two discs

Fig. B. 1 Metallic disc types.

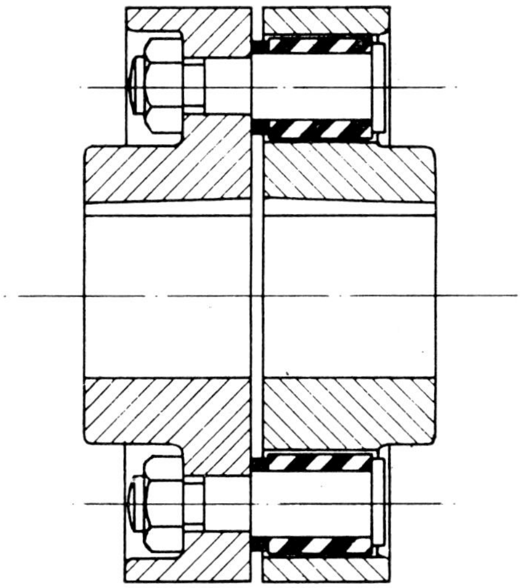

Resilient bushes loose fit in one flange

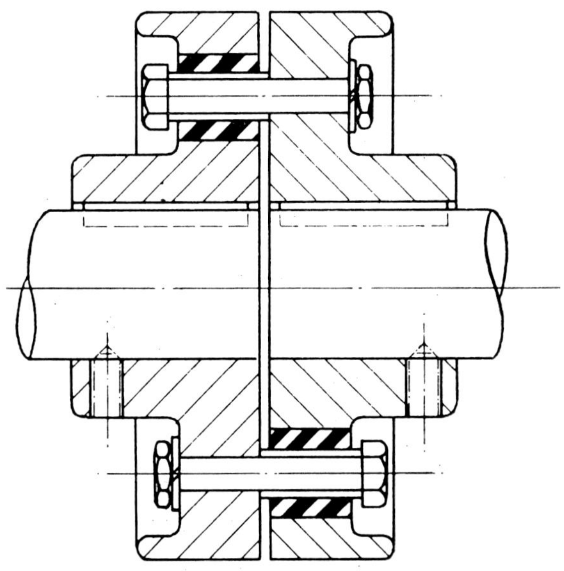

A design using bonded rubber bushes

Fig. B. 2 Pin types.

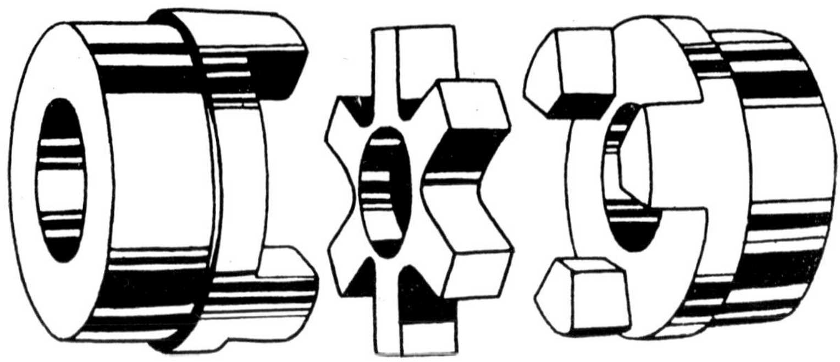

Spider coupling

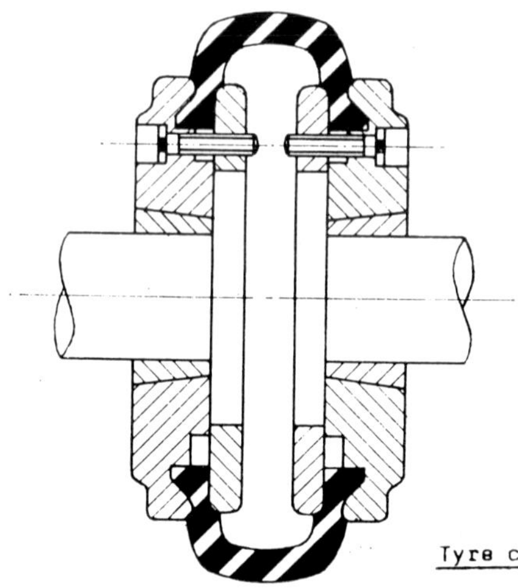

Tyre coupling

Fig. B. 3 Tyre and spider types

APPENDIX C

CONVERSION DATA

°F	Temperature (0° to 50°)	°C
32·0	0	−17·8
33·8	1	−17·2
35·6	2	−16·7
37·4	3	−16·1
39·2	4	−15·6
41·0	5	−15·0
42·8	6	−14·4
44·6	7	−13·9
46·4	8	−13·3
48·2	9	−12·8
50·0	10	−12·2
51·8	11	−11·7
53·6	12	−11·1
55·4	13	−10·6
57·2	14	−10·0
59·0	15	− 9·4
60·8	16	− 8·9
62·6	17	− 8·3
64·4	18	− 7·8
66·2	19	− 7·2
68·0	20	− 6·7
69·8	21	− 6·1
71·6	22	− 5·6
73·4	23	− 5·0
75·2	24	− 4·4
77·0	25	− 3·9
78·8	26	− 3·3
80·6	27	− 2·8
82·4	28	− 2·2
84·2	29	− 1·7
86·0	30	− 1·1
87·8	31	− 0·6
89·6	32	0·0
91·4	33	0·6
93·2	34	1·1
95·0	35	1·7
96·8	36	2·2
98·6	37	2·8
100·4	38	3·3
102·2	39	3·9
104·0	40	4·4
105·8	41	5·0
107·6	42	5·6
109·4	43	6·1
111·2	44	6·7
113·0	45	7·2
114·8	46	7·8
116·6	47	8·3
118·4	48	8·9
120·2	49	9·4
122·0	50	10·0

°F	Temperature (51° to 100°)	°C
123·8	51	10·6
125·6	52	11·1
127·4	53	11·7
129·2	54	12·2
131·0	55	12·8
132·8	56	13·3
134·6	57	13·9
136·4	58	14·4
138·2	59	15·0
140·0	60	15·6
140·0	60	15·6
141·8	61	16·1
143·6	62	16·7
145·4	63	17·2
147·2	64	17·8
149·0	65	18·3
150·8	66	18·9
152·6	67	19·4
154·4	68	20·0
156·2	69	20·6
158·0	70	21·1
159·8	71	21·7
161·6	72	22·2
163·4	73	22·8
165·2	74	23·3
167·0	75	23·9
168·8	76	24·4
170·6	77	25·0
172·4	78	25·6
174·2	79	26·1
176·0	80	26·7
177·8	81	27·2
179·6	82	27·8
181·4	83	28·3
183·2	84	28·9
185·0	85	29·4
186·8	86	30·0
188·6	87	30·6
190·4	88	31·1
192·2	89	31·7
194·0	90	32·2
195·8	91	32·8
197·6	92	33·3
199·4	93	33·9
201·2	94	34·4
203·0	95	35·0
204·8	96	35·6
206·6	97	36·1
208·4	98	36·7
210·2	99	37·2
212·0	100	37·8

Conversion Data (continued)

Pipe length
(feet and metres)
1 ft=0·304700 m
1 m=3·28084 ft

ft		m
3·28084	1	0·304800
6·56168	2	0·609600
9·84252	3	0·914400
13·1234	4	1·21920
16·4042	5	1·52400
19·6850	6	1·82880
22·9659	7	2·13360
26·2467	8	2·43840
29·5276	9	2·74320
32·8084	10	3·04800
36·0892	11	3·35280
39·3701	12	3·65760
42·6509	13	3·96240
45·9318	14	4·26720
49·2126	15	4·52700
52·4934	16	4·87680
55·7743	17	5·18160
59·0551	18	5·48640
62·3360	19	5·79120
65·6168	20	6·09600
68·8976	21	6·40080
72·1785	22	6·70560
75·4593	23	7·01040
78·7402	24	7·31520
82·0210	25	7·62000
98·4252	30	9·14400
114·829	35	10·6680
131·234	40	12·1920
147·638	45	13·7160
164·042	50	15·2400
180·446	55	16·7640
196·850	60	18·2880
213·255	65	19·8120
229·659	70	21·3360
246·063	75	22·8600
262·467	80	24·3840
278·871	85	25·9080
295·276	90	27·4320
311·680	95	28·9560
328·084	100	30·4800

Pipe bore
(inches and millimetres)
1 inch=25·4000 mm
1 mm=0·039370 in

in		mm
0·039370	1	25·4000
0·078740	2	50·8000
0·118110	3	76·2000
0·157480	4	101·600
0·196851	5	127·000
0·236221	6	152·400
0·275591	7	177·800
0·314961	8	203·200
0·354331	9	228·600
0·393701	10	254·000
0·433071	11	279·400
0·472441	12	304·800
0·511811	13	330·200
0·551181	14	355·600
0·590552	15	381·000
0·629922	16	406·400
0·669292	17	431·800
0·708662	18	457·200
0·748032	19	482·600
0·787402	20	508·000
0·826772	21	533·400
0·866142	22	558·800
0·905512	23	584·200
0·944882	24	609·600
0·984253	25	635·000
1·18110	30	762·000
1·37795	35	889·000
1·57480	40	1016·00
1·77165	45	1143·00
1·96851	50	1270·00
2·16536	55	1397·00
2·36221	60	1524·00
2·55906	65	1651·00
2·75591	70	1778·00
2·95276	75	1905·00
3·14961	80	2032·00
3·34646	85	2159·00
3·54331	90	2286·00
3·74016	95	2413·00
3·93701	100	2540·00

Conversion Data (continued)

Power
(horsepower and kilowatts)
1 hp = 0·745700 kW
1 kW = 1·34102 hp

hp		kW
1·34102	1	0·745700
2·68204	2	1·49140
4·02306	3	2·23710
5·36408	4	2·98280
6·70510	5	3·72850
8·04612	6	4·47420
9·38714	7	5·21990
10·7282	8	5·96560
12·0692	9	6·71130
13·4102	10	7·45700
14·7512	11	8·20270
16·0922	12	8·94840
17·4333	13	9·69410
18·7743	14	10·4398
20·1153	15	11·1855
21·4563	16	11·9312
22·7973	17	12·6769
24·1384	18	13·4226
25·4794	19	14·1683
26·8204	20	14·9140
28·1614	21	15·6597
29·5024	22	16·4054
30·8435	23	17·1511
32·1845	24	17·8968
33·5255	25	18·6425
40·2306	30	22·3710
46·9357	35	26·0995
53·6408	40	29·8280
60·3459	45	33·5565
67·0510	50	37·2850
73·7561	55	41·0135
80·4612	60	44·7420
87·1663	65	48·4705
93·8714	70	52·1990
100·577	75	55·9275
107·282	80	59·6560
113·987	85	63·3845
120·692	90	67·1130
127·397	95	70·8415
134·102	100	74·5700

Heat flow rate
(British thermal units per hour and watts)
1 Btu/h = 0·293071 W
1 W = 3·41214 Btu/h hp

Btu/h		W
3·41214	1	0·293071
6·82428	2	0·586142
10·2364	3	0·879213
13·6486	4	1·17228
17·0607	5	1·46536
20·4728	6	1·75843
23·8850	7	2·05150
27·2971	8	2·34457
30·7093	9	2·63764
34·1214	10	2·93071
37·5335	11	3·22378
40·9457	12	3·51685
44·3578	13	3·80992
47·7700	14	4·10299
51·1821	15	4·39607
54·5942	16	4·68914
58·0064	17	4·98221
61·4185	18	5·27528
64·8307	19	5·56835
68·2428	20	5·86142
71·6549	21	6·15449
75·0671	22	6·44756
78·4792	23	6·74063
81·8914	24	7·03370
85·3035	25	7·32678
102·364	30	8·79213
119·425	35	10·2575
136·486	40	11·7228
153·546	45	13·1882
170·607	50	14·6536
187·668	55	16·1189
204·728	60	17·5843
221·789	65	19·0496
238·850	70	20·5150
255·911	75	21·9803
272·971	80	23·4457
290·032	85	24·9110
307·093	90	26·3764
324·153	95	27·8417
341·214	100	29·3071

CONVERSION DATA (continued)

Mass
(pounds and kilogrammes)
1 lb=0·453592 kg
1 kg=2·20462 lb

lb		kg
2·20462	1	0·453592
4·40924	2	0·907184
6·61386	3	1·36078
8·81848	4	1·81437
11·0231	5	2·26796
13·2277	6	2·72155
15·4323	7	3·17515
17·6370	8	3·62874
19·8416	9	4·08233
22·0462	10	4·53592
24·2508	11	4·98952
26·4554	12	5·44311
28·6601	13	5·89670
30·8647	14	6·35029
33·0693	15	6·80389
35·2739	16	7·25748
37·4785	17	7·71107
39·6832	18	8·16466
41·8878	19	8·61826
44·0924	20	9·07185
46·2970	21	9·52544
48·5016	22	9·97903
50·7063	23	10·4326
52·9109	24	10·8862
55·1155	25	11·3398
66·1386	30	13·6078
77·1617	35	15·8757
88·1848	40	18·1437
99·2079	45	20·4117
110·231	50	22·6796
121·254	55	24·9476
132·277	60	27·2155
143·300	65	29·4835
154·323	70	31·7515
165·347	75	34·0194
176·370	80	36·2874
187·393	85	38·5554
198·416	90	40·8233
209·439	95	43·0913
220·462	100	45·3592

Flow rate
(gallons per min and cubic metres per hour)
1 gpm=0·27276 m³h
1 m³h=3·66615 gpm

gpm		m³h
3·66615	1	0·27276
7·33230	2	0·54552
10·99845	3	0·81828
14·66460	4	1·09104
18·33075	5	1·36380
21·99690	6	1·63656
25·66305	7	1·90932
29·32920	8	2·18208
32·99535	9	2·45484
36·66150	10	2·72760
40·32765	11	3·00036
43·99380	12	3·27312
47·65995	13	3·54588
51·32610	14	3·81864
54·99225	15	4·09140
58·65840	16	3·46416
62·32455	17	4·63692
65·94420	18	4·90968
69·65685	19	5·18244
73·32300	20	5·45520
76·98915	21	5·72796
80·65530	22	6·00072
84·32145	23	6·27348
87·98760	24	6·54624
91·65375	25	6·81900
109·98450	30	8·18280
128·31525	35	9·54660
146·64600	40	10·91040
164·97675	45	12·27420
183·30750	50	13·63800
201·63825	55	15·00180
219·96900	60	16·36560
238·29975	65	17·72940
256·63050	70	19·09320
274·96125	75	20·45700
293·29100	80	21·82080
311·62275	85	23·18460
329·95350	90	24·54840
348·28425	95	25·91220
366·61500	100	27·27600

Specific Gravity of Various Liquids

Liquid	Specific gravity	Reference temperature
Heavy Crude Oil	0·97	15°C
Castor Oil	0·969	15°C
Glycerol	1·2613	20°C
Olive Oil	0·918	15°C
Linseed Oil	0·942	15°C
Dowtherm 'A'	0·997	20°C
Kerosene	0·815	15°C
Turpentine	0·873	16°C
Mercury	13·546	20°C
Ethyl Alcohol	0·7893	20°C
Water	1·000	3·98°C
Benzene	0·879	20°C
Octane	0·7025	20°C
Heptane	0·6837	20°C
Carbon Disulphide	1·263	20°C
Hexane	0·6595	20°C

Approximate Imperial/Metric (SI) Equivalents

1 in.	=	25·4 mm
1 in^2	=	645·16 mm^2
1 UK Gal	=	4·546 dm^3
1 US Gal	=	3·785 dm^3
1 ft/s	=	0·3048 m/s
1 lb	=	0·4536 kg
10^5 N/m^2	=	1 bar
1 Std. Atmosphere	=	1·013 bar
1 lbf/in^2	=	6·892 kN/m^2
	=	$6{\cdot}892 \times 10^{-2}$ bar = 68·92 mbar
3 lbf/in^2	≈	20 kN/m^2 = 0·2 bar = 200 mbar
15 lbf/in^2	≈	0·1 MN/m^2 = 1·0 bar = 1000 mbar (≈1 atmos)
100 lbf/in^2	≈	0·7 MN/m^2 = 7 bar
145 lbf/in^2	≈	1·0 MN/m^2 = 10 bar
300 lbf/in^2	≈	2·0 MN/m^2 = 20 bar
1 hp	≈	746 Watt

BIBLIOGRAPHY

General References

ACKER, 'The Resistance of Fluids flowing in channels and pipes.' Paper No. 1 DSIR 1958 HMSO.

ADDISON, H., *Centrifugal and Other Rotodynamic Pumps* (Chapman and Hall).

ADDISON, H., *A Treatise on Applied Hydraulics* (Chapman and Hall).

KELLY, A. G., *Hydraulic Design of Syphons* (1. Mech. E., 65–66, vol. 180, part 1).

Pumping Manual, 2nd Edition by Trade & Technical Press Ltd.

LAZARKIEWIEZ, STEPHEN AND TROSKOLANSKI, ADAM T., *Impeller Pumps* (Pergamon).

PARMAKIAN, J., *Water Hammer Analysis* (Dover, New York).

STEPANOFF, A. J., *Centrifugal and Axial Flow Pumps* (Wiley, New York).

WILSON, Sir Alan, 'Final report of the committee on the problem of noise' (HMSO 1963 Command 2056).

E.E.U.A. Handbooks

No. 6: *Chemical Resistant Linings for Pipes and Vessels.*
18: *Supporting of Mild Steel and Cast Iron Pipes.*
20: *Thermoplastics Pipework and Ducting in Unplasticised Polyvinyl Chloride and Polythene.*
21: *Chemical Resistant Materials,* Part 1 and Part 2.
25: *Measurement and Control of Noise.*
26: *Electrically Driven Glandless Pumps.*
27: *Dynamic Sealing of Rotating and Reciprocating Shafts.*
29: *Electrical Variable Speed Drives.*
32: *The Selection of Automatic Control Valves.*

British Standards

B.S. 599: Methods of testing pumps.
1042: Code for flow measurement.
1394: Power driven circulators for heating plants.
1501–6: Steel for use in the chemical, petroleum and allied industries.
2613: Electrical performance of rotating electrical machines.
2765: Dimensions of temperature detecting elements and corresponding pockets.
2896: Built-on coolant pumps for internal combustion engines.
2993: Coolant pumps for internal combustion engines (nominal output up to 2500 gallons per hour), housing dimensions for face-type seals, nominal shaft diameters and suction and discharge stubs.
3766: Machine tool motorised centrifugal coolant pumps.
3829: Principal external dimensions of centrifugal pumps.
Part 1 – Horizontally split casing, double inlet pumps.
Part 2 – End suction pumps.
4082: External dimensions for vertical in-line centrifugal pumps.
Part 1 – 'I' Type.
Part 2 – 'U' type.
4519: External dimensions for horizontal end-suction centrifugal pumps
Part 1 : Medium duty – chemical.